T0332439

Modern Metaheuristics in Image Processing

Diego Oliva
Noe Ortega-Sánchez
Salvador Hinojosa
Marco Pérez-Cisneros

Universidad de Guadalajara
Guadalajara, Jalisco, México

CRC Press is an imprint of the
Taylor & Francis Group, an **informa** business
A SCIENCE PUBLISHERS BOOK

First edition published 2022
by CRC Press
6000 Broken Sound Parkway NW, Suite 300, Boca Raton, FL 33487-2742

and by CRC Press
4 Park Square, Milton Park, Abingdon, Oxon, OX14 4RN

CRC Press is an imprint of Taylor & Francis Group, LLC

Library of Congress Cataloging-in-Publication Data (applied for)

ISBN: 978-1-032-01977-2 (hbk)
ISBN: 978-1-032-02473-8 (pbk)
ISBN: 978-1-003-18350-1 (ebk)

DOI: 10.1201/9781003183501

Typeset in Times New Roman
by Radiant Productions

Preface

Metaheuristic algorithms are essential tools for solving different problems. They have been used to handle medicine, engineering, and image processing issues over the years. Several metaheuristics are proposed every year. Some of them have more sophisticated operators that permit a better performance in searching for the optimal solutions in complex search spaces.

This book presents a study of metaheuristic algorithms in image processing, such methods have been recently proposed, and they can be considered modern algorithms. The metaheuristic algorithms were selected based on their importance and use since they were initially proposed. All of them were published in 2020, and their use was extended to different areas of application. Since image processing and in specific image segmentation are critical parts of a computer vision system, they are different complexities and problems that could be solved using metaheuristics. Chapter 1 provides the basics of image processing and metaheuristic algorithms; the impact of both topics in the related literature is also analyzed. Chapter 2 presents a literature review of image segmentation and thresholding, which are important tasks in image processing. Chapter 3 introduces the use of the Political Optimizer for image thresholding by using the Otus's between-class variance. Chapter 4 explains the Manta Ray Foraging Optimization for multilevel thresholding. In this case, it is considered the Kapur's entropy as a criterion to find the best thresholds. In Chapter 5, the Archimedes Optimization

Algorithm is combined with the minimum cross entropy for image segmentation. Chapter 6 explains the basic concepts of the Equilibrium Optimizer and how it is combined with the Masi entropy for image thresholding. Finally, the MATLAB® codes for the algorithms used in this book are introduced and explained in Chapter 7.

The book has been written to be used in academic courses for undergraduate and postgraduate students in the fields of artificial intelligence or image processing. The aim is to show how it is possible to adapt the metaheuristic to solve problems in image processing. With the inclusion of different problems, the reader could practice with various methodologies such as Otsu's between class variance, Kapur's entropy, Minimum cross entropy, and Masi entropy. Besides, understanding how to modify the metaheuristics opens a vast range of possibilities with applications in biomedicine, surveillance, and industry.

Contents

Chapter 1
Introduction

In recent years the use of cameras has become popular in almost all aspects of our life. Currently, we have more than one camera embedded in our mobile phones, cars have cameras too, and in medicine, the use of them helps to diagnose different diseases. Cameras are also used in areas such as agriculture, surveillance, manufacturing, etc. Besides, due to technological advances, different kinds of cameras and devices to acquire images are available in the market. For example, thermal, ultrasonic, and hyperspectral cameras. Figure 1.1 shows an example of thermal images. Since they can be integrated in multiple devices their applications are infinite.

The use of cameras requires software which can process the information acquired by the sensor. It is called a computer vision

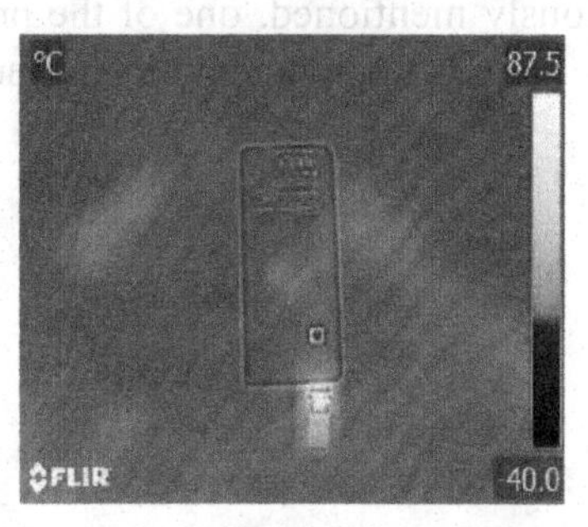

Fig. 1.1. An example of thermal images.

system, in it a sensing device comprises an input and a computer that interprets the input, and the output (Gonzalez and Woods, 1992). Figure 1.2 presents a simple diagram of a computer vision system. This kind of system is composed of different processing algorithms that permit the analysis of the features of the image. These methods are called image processing algorithms. Various approaches have been proposed in this area; for example, we can find algorithms to detect contours or corners (Erik Cuevas et al., 2011). Also, more complex approaches are used, for example, to identify the texture of the elements in the image. Nonetheless, one of the principal mechanisms from a computer vision system is segmentation (Singh et al., 2021). With the extensive use of cameras, more efficient methods are required to process the images; in this sense, researchers introduce the use of artificial intelligence (AI) algorithms.

The use of AI tools to improve image processing tasks permits more efficient algorithms that automatically analyze the image. Different methods as Neural Networks, Fuzzy sets, Metaheuristics, and Machine learning have been applied in image processing. Table 1.1 provides some examples of the approaches based on AI for solving image processing problems.

It is possible to see the use of AI algorithms in image processing to solve problems from different domains. This fact also permits a better interpretation of the scenes reflected in the accuracy that is expected in applications such as medical image processing.

As was previously mentioned, one of the primary tasks in a computer vision system is the segmentation of the images. Why is

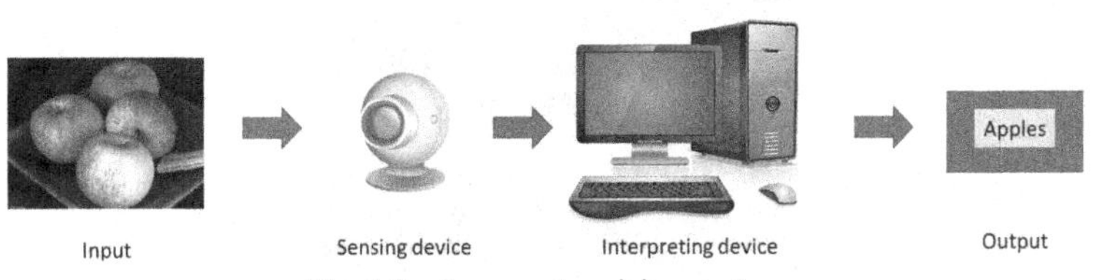

Fig. 1.2. A computer vision system.

Table 1.1. Some artificial intelligence methods in image processing.

AI technique	Application in image processing	References
Metaheuristics	Circle detection	(Cuevas et al., 2012)
	Template matching	(Oliva et al., 2014)
	Motion estimation	(Cuevas et al., 2013)
	Halftoning	(Ortega-Sánchez et al., 2020)
	Contrast enhancement	(Guha et al., 2021)
Bayesian networks	Segmentation	(Oliva et al., 2020)
Fuzzy logic		(Abd Elaziz et al., 2020)
	Corner detection	(Erik Cuevas et al., 2011)
Neural networks/ Deep Learning	Classification	(Ayaz et al., 2021)
		(Amin et al., 2018)
	Object recognition	(Pechebovicz et al., 2020)
Machine learning	Pixel's classification	(Guerrero et al., 2012)
	Object recognition/Classification	(Rodríguez-Esparza et al., 2020)
Hybrid methods	Classification	(Anter et al., 2021)

this task essential? To answer this question, we need to think about how our brains separate the object and therefore segmentation is crucial. It helps to separate objects that, in posterior steps, could be identified by other methods. In computational terms, segmentation makes it possible to separate the elements in a scene, for example, foreground and background. This is a simple way to use the segmentation, but when selecting more than one element in the scene, there are necessarily more complex algorithms. Figure 1.3(a) presents a plot of how the publication of articles in journals and conferences has been increasing between 2010 and 2020. Figure 1.3(b) shows the documents by area published in the same range of years.

In the related literature, it is possible to find different methodologies for segmentation. However, one of the easiest ways is to classify the pixels according to their intensity values. Under

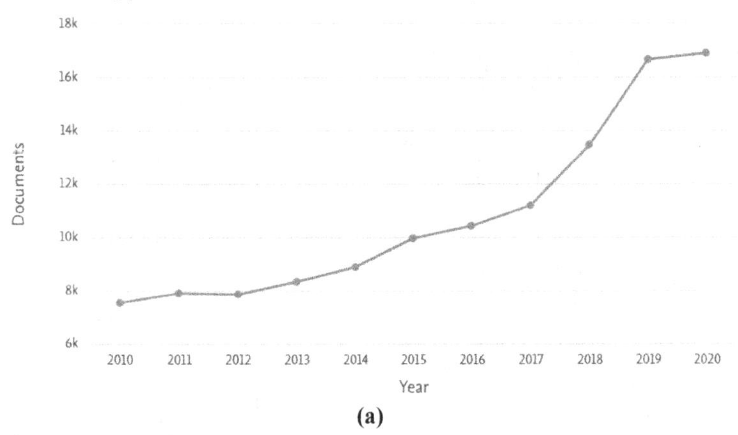

(a)

Documents by subject area

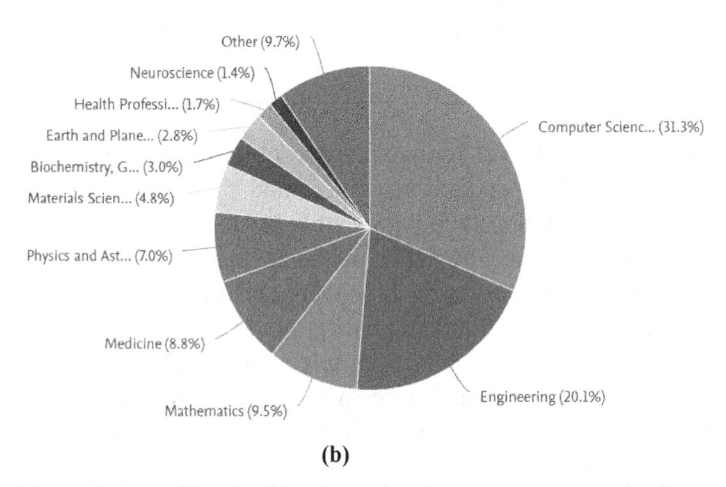

(b)

Fig. 1.3. (a) Relationship of publications related to image segmentation from 2010 to 2020. (b) Relationship of publications related to image segmentation by subject between the years 2010 to 2020 (Source Scopus).

such circumstances, a histogram is a tool that graphically the frequency of the intensities in the image. Besides, the histogram permits an analysis of the information contained in the scene. For example, we can see if an image has less or more illumination. Based on the histogram then is possible to start solving different problems in image processing. For image segmentation, the histogram permits the generation of groups of pixels by using a threshold value. Using a threshold value, it is possible to generate two groups of pixels and an output image where the background and foreground are identified. This process is called bi-level thresholding. When it is required to use more thresholds, it is called multilevel thresholding, and it helps to separate and define more objects in the scene.

1.1 Image Segmentation by Thresholding

The segmentation using thresholds is straightforward when the histogram is graphically available. However, the computers cannot see what we see in the histogram, and the appropriate selection of the thresholds becomes a problem. To handle this situation, different tools from statistics are used as the between-class variance proposed by Otsu (Otsu, 1979), the different kinds of entropies as presented by Kapur (Kapur et al., 1985), the Tsallis entropy (Oliva et al., 2019; Portes de Albuquerque et al., 2004), the minimum cross entropy (Li and Lee, 1993), and the Masi entropy (Masi, 2005). All these methods use the information of the histogram to select the optimal thresholds. For a single threshold, an exhaustive search is easy to implement since only two classes are generated. However, the computational effort increases when two or more thresholds are required (Diego Oliva et al., 2019). As shown in Figure 1.4, the multilevel thresholding produces better-segmented images compared to the bilevel approaches.

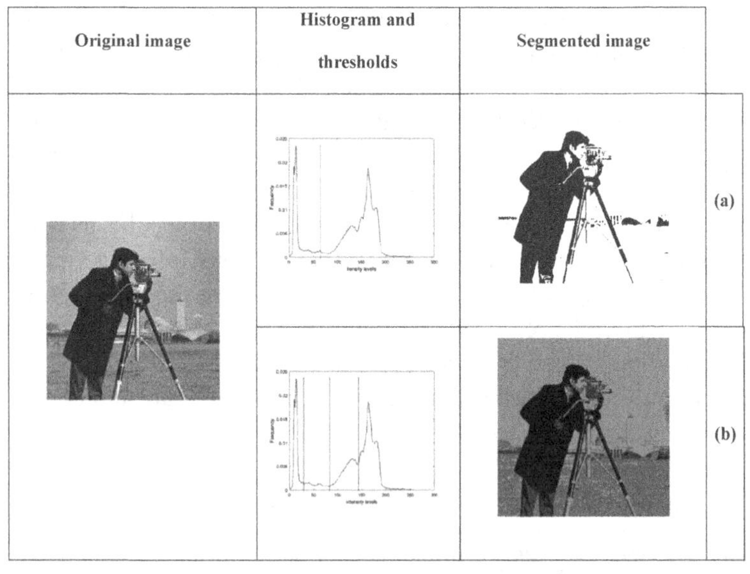

Fig. 1.4. Examples of segmentation for (a) bilevel thresholding and (b) multilevel thresholding.

More efficient methods are necessary to find the proper thresholds (Diego Oliva et al., 2019).

As was previously mentioned, thresholding is the easiest way to segment a digital image. In practical terms, the input grayscale image I_{Gr} has $m \times n$ pixels that will be divided using a threshold value th and the histogram with $L = 256$ intensity levels for an 8-bit digital image. The following equation explains the process for bilevel thresholding:

$$I_s(r,c) = \begin{cases} 0 & \text{if} \quad I_{Gr}(r,c) \leq th \\ L & \text{if} \quad I_{Gr}(r,c) > th \end{cases} \tag{1.1}$$

From Eq. 1.1, $I_{Gr}(r, c)$ is a gray value of the original image and $I_S(r, c)$ is a segmented value in the output image. The rule described above could be extended for more thresholds as follows:

$$I_s(r,c) = \begin{cases} I_{Gr}(r,c) & \text{if} \quad I_{Gr}(r,c) \leq th_1 \\ th_{i-1} & \text{if} \quad th_{i-1} < I_{Gr}(r,c) \leq th_i, \quad i = 2,3,\ldots nt-1 \\ I_{Gr}(r,c) & \text{if} \quad I_{Gr}(r,c) > th_{nt} \end{cases} \tag{1.2}$$

In Eq. 1.2, nt is the total number of thresholds that will create nt+1 classes; r and c correspond to a row and column position in the image.

1.1.1 Image segmentation quality metrics

Image thresholding is a non-supervised segmentation method that does not require any ground truth to verify its performance. Basically, it only needs the number of thresholds and the histogram of the image. However, to verify the quality of the segmentation, the segmented images need to be in order. Since images are signals, it is common to use metrics from digital signal processing. Table 1.2 presents the quality metrics most used in image thresholding. However, among them, the Peak Signal to Noise Ratio (PSNR) (Sankur et al., 2002), the Structural Similarity Index (SSIM), and the Feature Similarity Index (FSIM) are the most popular in the related literature (Diego Oliva et al., 2019; Varnan et al., 2011).

1.2 Metaheuristic Algorithms

One crucial part of AI is using optimization algorithms to search for the best solutions to complex problems that can be seen as black boxes. Metaheuristic algorithms (MAs) are optimization tools that employ different operators to explore and exploit the different regions of a search space. An optimization problem then must be adequately defined as follows (Eq. 1.3):

$$\begin{aligned} &\text{max/min} \quad f(\mathbf{x}), \ \mathbf{x} = (x_1, x_2, \ldots, x_d) \in \mathbb{R}^d \\ &\text{Subject to:} \quad l_i \leq x_i \leq u_i, i = 1,2,\ldots d \end{aligned} \tag{1.3}$$

Table 1.2. Quality metrics for image segmentation.

No.	Metric	Formulation	Remark
1.	Peak Signal-to-Noise Ratio (PSNR) (Horé and Ziou, 2010)	$PSNR = 20\log_{10}\left(\frac{MAX_I}{\sqrt{MSE}}\right)$	Determines the ratio among the maximum possible power of a signal and the power of corruptive noise.
2.	Root Mean Square Error (RMSE) (Varnan et al., 2011)	$RMSE = \sqrt{\frac{1}{n}\sum_{j=1}^{n}\left(y_j - \hat{y}_j\right)^2}$	Estimates the average magnitude of the error between a predicted value and an actual value.
3.	Structural Similarity Index (SSIM) (Wang et al., 2004)	$SSIM = \frac{\left(2\mu_{I_r}\mu_{I_s} + C_1\right)\left(2\sigma_{I_rI_s} + C_2\right)}{\left(\mu_{I_r}^2 + \mu_{I_s}^2 + C_1\right)\left(\sigma_{I_r}^2 + \sigma_{I_s}^2 + C_2\right)}$	Measures the similarity of structural information between the original and the processed image.
4.	Feature Similarity Index (FSIM) (Lin Zhang et al., 2011)	$FSIM = \frac{\sum_{x\in\Omega} S_L(X)PC_m(X)}{\sum_{x\in\Omega} PC_m(X)}$	Calculates the phase congruency and the gradient magnitude to characterize the local quality of the image.
5.	Normalized Cross-Correlation (NCC) (Memon et al., 2016)	$NCC = \frac{1}{n}\sum_{x,y}\frac{1}{\sigma_f\sigma_t}f(x,y)t(x,y)$	Determines the similarity among two digital images.
6.	Average Difference (AD) (Memon et al., 2016; Varnan et al., 2011)	$AD = \frac{1}{MN}\sum_{i=1}^{M}\sum_{j=1}^{N}(x(i,j) - y(i,j))$	Assesses the average difference among a reference image and a test image.
7.	Maximum Difference (MD) (Memon et al., 2016; Varnan et al., 2011)	$MD = MAX\lvert x(i,j) - y(i,j)\rvert$	Appraises the maximum error signal.
8.	Normalized Absolute Error (NAE) (Memon et al., 2016)	$NAE = \frac{\sum_{i=1}^{M}\sum_{j=1}^{N}\lvert X(i,j) - X(i,j)\rvert}{\sum_{i=1}^{M}\sum_{j=1}^{N}\lvert X(i,j)\rvert}$	The NAE evaluates the absolute difference between the initial image and the resultant image.

where **x** is a candidate solution composed by different decision variables that are defined in the lower (l) an upper (u) limits of a d-dimensional search space.

As an example of an optimization function, in Eq. 1.4 is presented one of the benchmark problems used in optimization; it is called the Rastrigin function.

$$f(\mathbf{x}) = 10d + \sum_{i=1}^{d}\left[x_i^2 - 10\cos(2\pi x_i)\right] \tag{1.4}$$

The function is evaluated in the search space defined between the limits $x_i \in [-5.12, 5.12]$ for all $i = 1, 2, \ldots, d$. This function is highly multimodal but has the global minimum at $f(\mathbf{x}^*) = 0$ at $\mathbf{x}^* = [0,\ldots,0]$. The plot of the Rastrigin function in 2D is presented in Figure 1.5.

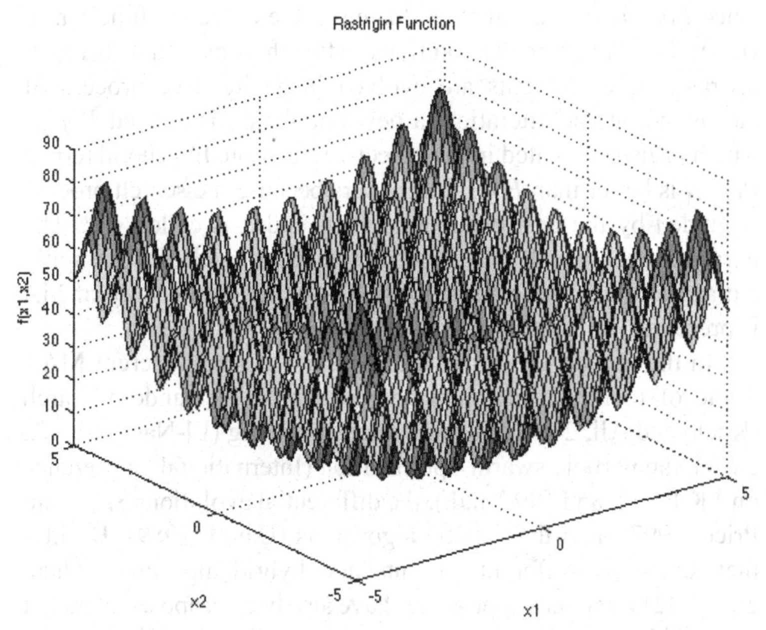

Fig. 1.5. The two-dimensional Rastrigin function.

1.2.1 A generic explanation of metaheuristic algorithms

To solve the problems defined in Eq. 1.3, the MAs are powerful alternatives since they do not require any gradient information to find the optimal solution. The benefits of MAs are the use of different rules that permit testing the candidate solutions in the objective function. Besides, by using various operators, the solutions are modified to guide the search to the most prominent regions in which there are more probabilities to find the global optimal.

An MA could work with a single solution or with a set of candidate solutions (called population). In practical terms, a population will be treated along with the book. In this sense, a population of N candidate solutions is defined as $Pop_p = (\mathbf{x}_1^p, \mathbf{x}_2^p, ..., \mathbf{x}_N^p)$. At the beginning of any MA, Pop_p is randomly initialized. Once Pop_p is created, it is evaluated in the objective function to verify the quality of the solutions. After that, by using different operators, the elements are evolved in an iterative process. It means that at each iteration, a new population is created Pop_{p+1} which is also evaluated in the objective function. In general terms, Pop_{p+1} is better than Pop_p; this occurs because the search process is guided by the best element of Pop_p and it is called g^{best}. The iterative process ends when a stop criterion is reached (commonly a maximum number of iterations). The generic procedure of MA is presented in Figure 1.6.

In the related literature, it is possible to find different MAs. Some of them are classical approaches as the random search (Kaelo and Ali, 2006), the simulated annealing (El-Naggar et al., 2012), the particle swarm optimization (International Conference on J.K.P. of I. and 1995, n.d.), the differential evolution (Storn and Price, 1997), and the genetic algorithms (David, 1989). Besides that, there are different variants and hybrid algorithms (Juan et al., 2021). Modern approaches have also been proposed in recent years. This book aims to explain how to implement different MAs recently proposed for solving the image thresholding problem.

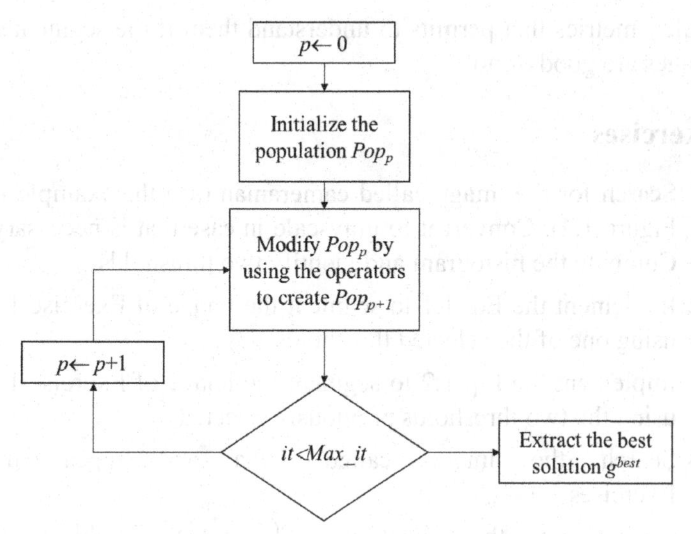

Fig. 1.6. Flowchart of a generic MA for optimization.

1.3 Conclusions

Image segmentation is an important field of research that still has open challenges and some of them could be addressed from an optimization point of view. This is not only for benchmark images but also for different fields of applications. On the other hand, the tendency of having new metaheuristics requires to understand them and see how they work. During the years, researchers have been implementing metaheuristics to solve segmentation problems, especially for thresholding. The problem of the modern algorithms is that the metaphor and the adaptation of the operators to follow the metaphor makes them complicated to understand and modify. This book aims to provide the reader with a guide that permits to understand how to convert a metaheuristic into a method for image thresholding by using the most popular mechanism based on the histogram. Besides that, the chapters include an analysis of the segmentation results by considering quality metrics that permits to understand them if the segmented images are good or not.

Exercises

1.1 Search for the image called cameraman (see the example in Figure 1.4). Convert it to grayscale in case that is necessary. Compute the histogram and identify two thresholds.

1.2 Implement the Eq. 1.1 to segment the image of Exercise 1.1 using one of the selected thresholds.

1.3 Implement the Eq. 1.2 to segment the image of Exercise 1.1 using the two thresholds previously selected.

1.4 Search the image called Lena and repeat the Exercises 1.1–1.3.

1.5 Considering the function: $f(\mathbf{x}) = \sum_{i=1}^{10} x_i^2$, subject to $x_i \in [-5.12, 5.12]$

Determine:

- The bounds of the search space
- The number of dimensions
- The decision variables.

References

Abd Elaziz, M., Sarkar, U., Nag, S., Hinojosa, S. and Oliva, D. (2020). Improving image thresholding by the type II fuzzy entropy and a hybrid optimization algorithm. *Soft Computing*, 24(19): 14885–14905. https://doi.org/10.1007/s00500-020-04842-7.

Amin, J., Sharif, M., Yasmin, M. and Fernandes, S.L. (2018). Big data analysis for brain tumor detection: Deep convolutional neural networks. *Future Generation Computer Systems*, 87: 290–297. https://doi.org/10.1016/j.future.2018.04.065.

Anter, A.M., Oliva, D., Thakare, A. and Zhang, Z. (2021). AFCM-LSMA: New intelligent model based on Lévy slime mould algorithm and adaptive fuzzy C-means for identification of COVID-19 infection from chest X-ray images. *Advanced Engineering Informatics*, 49(May): 101317. https://doi.org/10.1016/j.aei.2021.101317.

Ayaz, H., Rodríguez-Esparza, E., Ahmad, M., Oliva, D., Pérez-Cisneros, M. and Sarkar, R. (2021). Classification of apple disease based on non-linear deep features. *Applied Sciences* (*Switzerland*), 11(14). https://doi.org/10.3390/app11146422.

Cuevas, E., Zaldivar, D., Pérez-Cisneros, M., Sánchez, E. and Ramírez-Ortegón, M. (2011). Robust fuzzy corner detector. *Intelligent Automation and Soft Computing*, 17(4): 415–429. https://doi.org/10.1080/10798587.2011.10643158.

Cuevas, E, Oliva, D., Zaldivar, D., Pérez-Cisneros, M. and Sossa, H. (2012). Circle detection using electro-magnetism optimization. *Information Sciences*. https://doi.org/10.1016/j.ins.2010.12.024.

Cuevas, E., Zaldívar, D., Pérez-Cisneros, M. and Oliva, D. (2013). Block-matching algorithm based on differential evolution for motion estimation. *Engineering Applications of Artificial Intelligence*. https://doi.org/10.1016/j.engappai.2012.08.003.

David, G. (1989). *Genetic Algorithms in Search, Optimization, and Machine Learning* (1st Edn.). Boston, MA, USA: Addison-Wesley.

Eberhart, R. and Kennedy, J. (1995, November). Particle swarm optimization. pp. 1942–1948. In Proceedings of the IEEE International Conference on Neural Networks (Vol. 4).

El-Naggar, K.M., Al Rashidi, M.R., Al Hajri, M.F. and Al-Othman, A.K. (2012). Simulated Annealing algorithm for photovoltaic parameters identification. *Solar Energy*, 86(1): 266–274. https://doi.org/10.1016/j.solener.2011.09.032.

Gonzalez, R.C. and Woods, R.E. (1992). *Digital Image Processing*. New Jersey: Pearson, Prentice-Hall.

Guerrero, J.M., Pajares, G., Montalvo, M., Romeo, J. and Guijarro, M. (2012). Support Vector Machines for crop/weeds identification in maize fields. *Expert Systems with Applications*, 39(12): 11149–11155. https://doi.org/10.1016/j.eswa.2012.03.040.

Guha, R., Alam, I., Bera, S.K., Kumar, N. and Sarkar, R. (2021). Enhancement of image contrast using Selfish Herd Optimizer. *Multimedia Tools and Applications* (November 2020). https://doi.org/10.1007/s11042-021-11404-y.

Horé, A. and Ziou, D. (2010). Image quality metrics: PSNR vs. SSIM. *Proceedings - 20th International Conference on Pattern Recognition*, (March 2015): 2366–2369. https://doi.org/10.1109/ICPR.2010.579.

Juan, A.A., Keenan, P., Martí, R., McGarraghy, S., Panadero, J., Carroll, P. and Oliva, D. (2021). A review of the role of heuristics in stochastic optimisation: From metaheuristics to learnheuristics. *Annals of Operations Research*. https://doi.org/10.1007/s10479-021-04142-9.

Kaelo, P. and Ali, M.M. (2006). Some variants of the controlled random search algorithm for global optimization. *Journal of Optimization Theory and Applications*, 130(2): 253–264. https://doi.org/10.1007/s10957-006-9101-0.

Kapur, J.N., Sahoo, P.K. and Wong, A.K.C. (1985). A new method for gray-level picture thresholding using the entropy of the histogram. *Computer Vision, Graphics, and Image Processing*, 29(3): 273–285.

Li, C.H. and Lee, C.K. (1993). Minimum cross entropy thresholding. *Pattern Recognition*, 26(4): 617–625. https://doi.org/10.1016/0031-3203(93)90115-D.

Lin Zhang, Lei Zhang, Xuanqin Mou and Zhang, D. (2011). FSIM: A feature similarity index for image quality assessment. *IEEE Transactions on Image Processing*, 20(8): 2378–2386. https://doi.org/10.1109/TIP.2011.2109730.

Masi, M. (2005). A step beyond Tsallis and Rényi entropies. *Physics Letters, Section A: General, Atomic and Solid State Physics*, 338(3–5): 217–224. https://doi.org/10.1016/j.physleta.2005.01.094.

Memon, F.A., Unar, M.A. and Memon, S. (2016). Image Quality Assessment for Performance Evaluation of Focus Measure Operators. *ArXiv*, *abs/1604.0*.

Oliva, D, Cuevas, E., Pajares, G. and Zaldivar, D. (2014). Template matching using an improved electromagnetism-like algorithm. *Applied Intelligence*. https://doi.org/10.1007/s10489-014-0552-y.

Oliva, D., Abd Elaziz, M. and Hinojosa, S. (2019). Tsallis entropy for image thresholding. *In*: *Studies in Computational Intelligence* (Vol. 825). https://doi.org/10.1007/978-3-030-12931-6_9.

Oliva, D., Martins, M.S.R., Osuna-Enciso, V. and de Morais, E.F. (2020). Combining information from thresholding techniques through an evolutionary Bayesian network algorithm. *Applied Soft Computing Journal*, 90. https://doi.org/10.1016/j.asoc.2020.106147.

Oliva, Diego, Elaziz, M.A. and Hinojosa, S. (2019). *Metaheuristic Algorithms for Image Segmentation: Theory and Applications*. Springer.

Ortega-Sánchez, N., Oliva, D., Cuevas, E., Pérez-Cisneros, M. and Juan, A.A. (2020). An evolutionary approach to improve the halftoning process. *Mathematics*, 8(9). https://doi.org/10.3390/MATH8091636.

Otsu, N. (1979). A threshold selection method from gray-level histograms. *IEEE Transactions on Systems, Man, and Cybernetics*, 9(1): 62–66. https://doi.org/10.1109/TSMC.1979.4310076.

Pechebovicz, D., Premebida, S., Soares, V., Camargo, T., Bittencourt, J.L., Baroncini, V. and Martins, M. (2020). Plants recognition using embedded convolutional neural networks on mobile devices. *Proceedings of the IEEE International Conference on Industrial Technology, 2020–February*, 674–679. https://doi.org/10.1109/ICIT45562.2020.9067289.

Portes de Albuquerque, M., Esquef, I.A., Gesualdi Mello, A.R. and Portes de Albuquerque, M. (2004). Image thresholding using Tsallis entropy. *Pattern Recognition Letters*, 25(9): 1059–1065. https://doi.org/10.1016/j.patrec.2004.03.003.

Rodríguez-Esparza, E., Zanella-Calzada, L., Oliva, D. and Perez-Cisneros, M. (2020). Automatic detection and classification of abnormal tissues on digital mammograms based on a bag-ofvisual-words approach. *Proc. SPIE 11314, Medical Imaging 2020: Computer-Aided Diagnosis*, 1131424(16). https://doi.org/10.1117/12.2549899.

Sankur, B., Sankur, B. and Sayood, K. (2002). Statistical evaluation of image quality measures. *Journal of Electronic Imaging*, 11(2): 206. https://doi.org/10.1117/1.1455011.

Singh, S., Mittal, N., Thakur, D., Singh, H., Oliva, D. and Demin, A. (2021). Nature and biologically inspired image segmentation techniques. *Archives of Computational Methods in Engineering* (0123456789). https://doi.org/10.1007/s11831-021-09619-1.

Storn, R. and Price, K. (1997). Differential evolution: A simple and efficient heuristic for global optimization over continuous spaces. *Journal of Global Optimization*, 11: 341–359.

Varnan, C.S., Jagan, A., Kaur, J., Jyoti, D. and Rao, D.S. (2011). Image quality assessment techniques in spatial domain. *International Journal of Computer Science and Technology*, 2(3): 177–184.

Wang, Z., Bovik, A.C., Sheikh, H.R. and Simoncelli, E.P. (2004). Image quality assessment: From error visibility to structural similarity. *IEEE Transactions on Image Processing*, 13(4): 600–612. https://doi.org/10.1109/TIP.2003.819861.

CHAPTER 2
Literature Review

2.1 Introduction

In this chapter, we will analyze the literature concerning image thresholding. First, we will discuss which nonparametric criteria have had a more significant impact in the area. Then, recent articles from the last five years are analyzed in terms of their contributions, including perspectives such as the multidimensional histograms, energy curve, multi-objective perspectives, etc.

2.2 Multilevel Thresholding over the Years

The problem of image thresholding can be traced back to more than 50 years ago when the first descriptions of statistical measures were used to determine if two sets of pixels were likely to come from the same distribution, known as Otsu's method (Otsu, 1979). Since then, many variations of statistical criteria have been proposed; the entropy-based formulations being the most prolific in terms of variations. The entropy-based appróaches started with the development by Kapur et al. (1985). Thereafter, other techniques emerged, such as the cross-entropy, Rényi and Tsallis enrtophy formulations (Li and Lee, 1993; Li et al., 2007; Sahoo and Arora, 2004). To this list recently, the Masi entropy has added (Khairuzzaman and Chaudhury, 2019).

In the literature it is hard to properly disseminate the impact of a specific topic. To visualize how each criterion has contributed to the area, we used Scopus to inspect how many articles have been published over the years, including the terms image thresholding and the name of the criterion. Figure 2.1 shows the results obtained by multiple queries up to the middle of 2021. Clearly, the most used approach is Otsu's method, as it is the oldest and is relatively easy to implement. The second approach is the Kapur formulation. The newer entropies typically compare their performance against Otsu or Kapur's methods; the occurrences of both algorithms on almost any thresholding article have helped them increase their numbers. The cross-entropy, Rényi, and Tsallis behave very similarly, while the Masi entropy has just started its path.

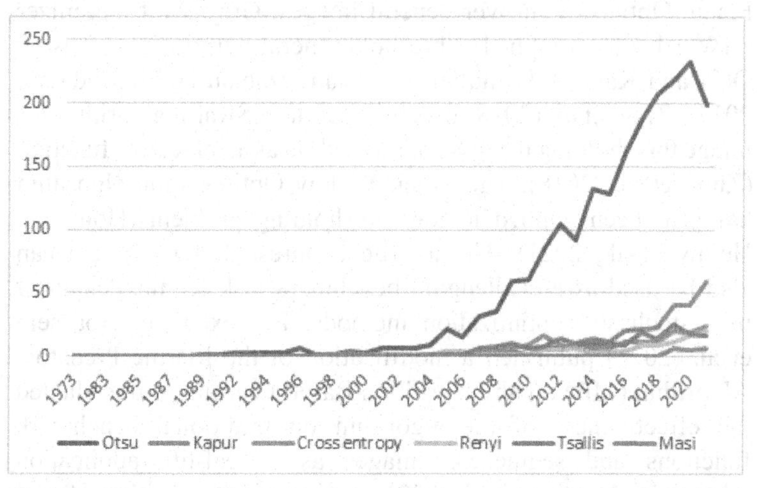

Fig. 2.1. Nonparametric approaches over the years applied to multilevel thresholding.

2.3 Current Trends on Image Thresholding

Over the last few years, many approaches have been considered for the multilevel thresholding problem. In this subsection, we will describe the most important trends on the topic of image thresholding. For the convenience of the reader, the analysis will be conducted considering the following issues: new metaheuristics and modification of algorithms, hyper-heuristic approaches, multi-objective implementations, multidimensional histogram formulations, and the energy curve formulation.

2.3.1 New metaheuristics and modifications

The most important trend on image thresholding is the evaluation of new approaches that can outperform state-of-the-art by using new stochastic optimization algorithms.

In 2017, the article by Aziz et al. (2017) analyzed the performance of the Whale Optimization Algorithm and the Moth-Flame Optimization over general images. Grey Wolf Optimizer (GWO) has been applied to threshold general-purpose images using Otsu and Kapur's formulations (Khairuzzaman and Chaudhury, 2017). Zhou et al. (2018) presented a Moth Swarm algorithm for image thresholding using Kapur's method as an objective function (Zhou et al., 2018). The Black Widow Optimization algorithm has also been applied to the thresholding problem (Houssein, Helmy et al., 2021). The multilevel thresholding problem can also be used as a challenging benchmark task for the design of new stochastic optimization methods. For example, Houssein et al. (2021) published a modification of the Marine Predators Algorithm (MPA) with Opposition Based Learning and evaluated the effectiveness of the algorithm on traditional benchmark functions and segmented images as a real-life application (Houssein, Hussain et al., 2021). Another hybridization of two metaheuristic algorithms mixes the volleyball premier league algorithm and whale optimization algorithm using Otsu's formulation as an objective function (Abd Elaziz et al., 2021). We can find an example of color image thresholding in the work of He et al. (2017) and Liang et al. (2019). Table 2.1 presents a list of the metaheuristic and improved version used in image segmentation.

Table 2.1. Examples of new implementations of metaheuristic algorithms and enhancements applied to multilevel thresholding.

Article	**Title**	**Metaheuristic**	**Objective Function**	**Metrics**	**Application**
(Aziz et al., 2017)	Whale optimization Algorithm and Moth-Flame Optimization for multilevel thresholding image segmentation	Whale Optimization Algorithm (WOA) and Moth-Flame Optimizer (MFO)	Otsu	PSNR, SSIM, time	General images
(Houssein, Helmy, et al., 2021)	A novel Black Widow Optimization algorithm for multilevel thresholding image segmentation	Black Widow Optimizer (BWO)	Otsu, Kapur	PSNR, SSIM, and FSIM	General images
(Zhou et al., 2018)	Metaheuristic moth swarm algorithm for multilevel thresholding image segmentation	Moth Swarm Algorithm (MSA)	Kapur	PSNR, SSIM, and time	General images
(Khairuzzaman and Chaudhury, 2017)	Multilevel thresholding using gray wolf optimizer for image segmentation	Gray Wolf Optimizer	Otsu, Kapur	Mean Structural SIMilarity (MSSIM)	General images
(Houssein, Hussain et al., 2021)	An improved opposition-based marine predators' algorithm for global optimization and multilevel thresholding image segmentation	Marine Predators Algorithm with Opposition-Based Learning (MPA-OBL)	Otsu and Kapur	PSNR, SSIM, FSIM	General images
(Yan et al., 2020)	Modified Water Wave optimization algorithm for underwater multilevel thresholding image segmentation	Water Wave Optimizer (WWO)	Kapur	PSNR, SSIM	Underwater images

Table 2.1 contd. ...

...Table 2.1 contd.

Article	Title	Metaheuristic	Objective Function	Metrics	Application
(Khairuzzaman and Chaudhury, 2019)	Masi entropy based multilevel thresholding for image segmentation	Particle Swarm Optimization (PSO)	Masi entropy	MSSIM, PSNR	General images
(Abd Elaziz et al., 2021)	Multilevel thresholding image segmentation based on improved Volleyball Premier League algorithm using Whale Optimization algorithm	Hybridization of Volleyball Premier League (VPL) with Whale Optimization Algorithm (WOA), VPLWOA	Otsu	PSNR, SSIM	General images
(Liang et al., 2019)	Modified Grasshopper Algorithm-Based Multilevel Thresholding for Color Image Segmentation	Grasshopper Optimization Algorithm (GOA)	Otsu and Rényi	PSNR	Color images
(He and Huang, 2017)	Modified Firefly Algorithm based multilevel thresholding for color image segmentation	Modified Firefly Algorithm (MFA)	Otsu, Kapur, cross entropy	PSNR, SSIM	Color images
(Yan et al., 2021)	Kapur's entropy for underwater multilevel thresholding image segmentation based on Whale Optimization Algorithm	Whale Optimization Algorithm (WOA)	Kapur	PSNR, SSIM	Underwater images

2.3.2 Hyper-heuristics

The topic of hyper-heuristics includes approaches that utilize a higher-level heuristic to select a low-level heuristic that can be applied for a specific problem. In the context of image thresholding, the hyper-heuristic-based methods will generate a particular heuristic designed for each image to be segmented. This kind of approach might fall into the last category, but we consider that there is great potential in this area to highlight it as a separate category. For example, we can observe the article proposed by Elaziz et al. (2019) where they present a Swarm Selection methodology that uses a Differential Evolution algorithm to select the best type of swarm-based algorithm that can be applied over a specific image. The study includes 10 popular algorithms working with Otsu as an objective function (Abd Elaziz et al., 2019). Also, the article by Elaziz et al. (2020) proposes a hyperheuristic framework based on the Genetic Algorithm (GA) designed to select the best combination of metaheuristics for a given set of images. Table 2.2 shows some examples of hyperheuristics used in multilevel thresholding.

Table 2.2. Examples of new implementations of hyperheuristics applied to multilevel thresholding.

Article	Title	Metaheuristic	Objective function	Metrics	Type of images
(Abd Elaziz et al., 2019)	Swarm Selection method for multilevel thresholding image segmentation	Swarm Selection (SS)	Otsu	PSNR, SSIM, time	General images
(Elaziz et al., 2020)	Hyperheuristic method for multilevel thresholding image segmentation	Hyperheuristic based on GA	Otsu	PSNR, SSIM and time	General images

2.3.3 Multi-objective thresholding

In chapter one, the problem of optimization was addressed as a single-objective problem, where an algorithm explores the search space to find the optimal combination of decision variables. In the context of multilevel thresholding, the problem is to minimize/maximize a nonparametric criterion while the algorithm searches for the best combination of decision variables, in this case, threshold values. Thus, we can easily compare two sets of thresholds to determine which one is better by comparing their numerical fitness values.

The problem of thresholding can be reformulated as a multi-objective problem, where two or more conflicting objective functions coexist during the optimization process. Multi-objective problems usually rely on a mechanism designed to compare two solutions in terms of different objectives through the Pareto front. The goal of a multi-objective optimizer is to find a set of solutions contained in the Pareto front instead of finding a single solution. The elements of the Pareto front are said to be better than the other feasible solutions that are not included in the Pareto front, while the elements of the Pareto front are better than the others, at least in one objective. We can find a clear example of multilevel thresholding from a multi-objective perspective in an article about unassisted thresholding (Hinojosa et al., 2018); there the authors proposed a segmentation process with two conflicting goals. The first objective is to enhance the quality of the segmentation by minimizing the cross-entropy, while the other objective is to minimize the number of thresholds. The objective is said to be opposed as if we reduce the number of thresholds and the quality of the segmentation reduces, so we cannot minimize those two approaches simultaneously. As a result, we need to find a good trade-off between objectives.

In Table 2.3, we can observe some recent and relevant approaches suited for image thresholding using multi-objective approaches. In the recent literature, we can find approaches that

Table 2.3. Examples of new implementations of multi-objective schemes applied to multilevel thresholding.

Article	Title	Metaheuristic	Fitness Function	Metrics	Type of images
(Yan et al., 2021)	Kapur's Entropy for underwater multilevel thresholding image segmentation based on Whale Optimization Algorithm	Whale Optimization Algorithm (WOA)	Kapur	PSNR, SSIM	Underwater images
(Xing and He, 2021)	Many-objectives multilevel thresholding image segmentation for infrared images of power equipment with boost marine predators' algorithm	Boost Marine Predator Algorithm (BMPA)	9-Dimensional Kapur	PSNR, FSIM, Hypervolume, Spacing	Infrared images
(Elaziz et al., 2019)	Multilevel thresholding-based grayscale image segmentation using Multi-Objective Multivverse Optimizer	Multi-Objective Multiverse Optimizer (MOMVO)	Otsu and Kapur	PSNR, SSIM, Hypervolume, Spacing	General images
(Hinojosa et al., 2020)	Reducing overlapped pixels: A multi-objective color thresholding approach	NSGA-III	Kapur	Overlapping Index, PSNR, FSIM, SSIM, Hypervolume	Color images
(Sarkar et al., 2017)	Multilevel thresholding with a decomposition-based multi-objective evolutionary algorithm for segmenting natural and medical images	MOEA/D-DE	Cross-entropy and Rényi Entropy	Uniformity, hypervolume, spacing	Brain images

Table 2.3 contd. ...

...Table 2.3 contd.

Article	Title	Metaheuristic	Fitness Function	Metrics	Type of images
(Karakoyun et al., 2021)	D-MOSG: Discrete multi-objective shuffled gray wolf optimizer for multilevel image thresholding	Discrete multi-objective shuffled gray wolf optimizer (D-MOSG)	Otsu and Kapur	PSNR, FSIM,	General images
(Hinojosa et al., 2018)	Unassisted thresholding based on multi-objective evolutionary algorithms	NSGA-III	Cross-entropy and number of thresholds	PSNR, FSIM, SSIM, Hypervolume	General images
(Wunnava et al., 2020a)	A novel interdependence based multilevel thresholding technique using Adaptive Equilibrium Optimizer	Adaptive Equilibrium Optimizer (AEO)	Interdependency	PSNR, SSIM, FSIM	General images
(Hinojosa et al., 2019)	Remote sensing imagery segmentation based on multi-objective optimization algorithms	NSGA-II	Cross-entropy over multiple channels	PSNR, SSIM	Remote sensing images
(Elaziz and Lu, 2019)	Many-objectives multilevel thresholding image segmentation using knee evolutionary algorithm	Knee Evolutionary Algorithm (KnEA)	7 objectives: Otsu, Kapur, Fuzzy entropy, Cross entropy, Tsallis entropy, Rényi's entropy, and Fuzzy C-Means.	PSNR, SSIM, computational time, Hypervolume and coverage	General images

use the multi-objective framework to simultaneously optimize different thresholding criteria where the work by Elasiz and Lu (2019) stands out by formulating a nine-dimensional problem with as many criteria (Elaziz et al., 2019; Elaziz and Lu, 2019; Karakoyun et al., 2021; Sarkar et al., 2017). From an evaluation perspective, multi-objective approaches also use the same metrics for image quality, but due to its nature, it is common to find metrics designed to evaluate the convergence of the algorithm, such as Hypervolume and spacing (Xing and He, 2021). In Elaziz and Lu (2019) the authors proposed an interesting multi-objective approach that considers seven objective functions to be optimized simultaneously by the Knee Evolutionary Algorithm and compared them against other well-known multi-objective approaches.

2.3.4 Multidimensional histograms

One of the main drawbacks of the multilevel thresholding approach is that its formulation is based on the histogram of the image. While histogram-based approaches are fast and easy to compute, the information of an image is projected into a curve over a single dimension, and much information is lost. Over the recent years, many approaches have tried to extend the abilities of multilevel thresholding by including multidimensional formulations of the histogram. In Table 2.4, we can observe some recent and relevant approaches suited for image thresholding using multidimensional histograms or entropies. In general, all the approaches discussed over this subsection take the histogram of the image and another type of information to generate an n-dimensional search space where a metaheuristic algorithm will hunt for an optimum configuration. We can observe that the Rényi entropy and Otsu-based modifications are the most common objective functions (Bhandari and Kumar, 2019; Kumar et al., 2021; Mittal and Saraswat, 2018; Vig and Kumar, 2021). Regarding the metaheuristic algorithms, there is not a single algorithm that is widespread over the others; however, many

Table 2.4. Examples of new implementations of multidimensional histograms applied to multilevel thresholding.

Article	Title	Metaheuristic	Fitness function	Metrics	Type of images
(Kumar et al., 2021)	A new multilevel histogram thresholding approach using variational mode decomposition	Multilevel iterative Variational Mode Decomposition (VMD)	Rényi	ME, NAE, MSE, PSNR, FSIM, SSIM	Color images
(Vig and Kumar, 2021)	Comparison of different metaheuristic algorithms for multilevel non-local means 2D histogram thresholding segmentation	Whale Optimization Algorithm (WOA)	Rényi	Fitness value	Color images
(Agrawal et al., 2020)	A novel diagonal class entropy-based multilevel image thresholding using Coral Reef Optimization	Coral Reef Optimization (CRO)	Diagonal Class Entropy (DCE)	PSRN, SSIM, FSIM, cross-correlation (CC)	General images
(Mittal and Saraswat, 2018)	An optimum multilevel image thresholding segmentation using non-local means 2D histogram and exponential Kbest gravitational search algorithm	Exponential Kbest gravitational search algorithm	2D non-local means and Rényi	12 performance metrics	Color images
(Borjigin and Sahoo, 2019)	Color image segmentation based on multilevel Tsallis–Havrda–Charvát entropy and 2D histogram using PSO algorithms	PSO	Generalized 2D multilevel thresholding criterion	BD, PRI, GCE, and VOI	Color images

(Wunnava et al., 2020b)	An adaptive Harris Hawks optimization technique for 2D gray gradient based multilevel image thresholding	Harris Hawks Optimizer (HHO)	Normalized gray gradient based 2D histogram	PSNR, FSIN, SSIM	Color images
(Bhandari and Kumar, 2019)	A context sensitive energy thresholding-based 3D Otsu function for image segmentation using Human Learning Optimization	Human Learning Optimization (HLO)	3D Otsu	MSE, PSNR, FSIM, SSIM	Color images

approaches include a variation over a previously published article (Mittal and Saraswat, 2018; Wunnava et al., 2020b). Although most approaches rely on the histogram of a given image, other approaches have taken another path. For instance, the histogram formulation was extended into a 2D representation for the Rényi and Tsallis entropies (Ben Ishak, 2017). Following a similar idea, Tan and Zhang presented a modified Gravitational Search Algorithm (GSA) that uses a fuzzy system to select the parameters of the GSA (Tan and Zhang, 2020). The approach works to find optimal thresholds over a 2D representation of the Tsallis entropy.

2.3.5 Energy curve

Another interesting approach that has been receiving increased attention over the past few years is the use of other types of information for the thresholding process. The most popular alternative is the energy curve. The formulation of the energy curve considers not only the intensity value of each pixel but also includes a spatial relationship within a vicinity. For example, we can observe in Table 2.5 that Srikanth and Bikshalu (2021) proposed a Harmony Search approach using the formulation of the energy curve (Srikanth and Bikshalu, 2021); such an article is interesting as it uses nontraditional quality metrics for the evaluation of the segmentation, such as Dunn Index, Davies-Bouldin Index, SD validity index, Probabilistic Rand Index.

2.4 Benchmark Images

One key element in the different research fields is the comparability and reproducibility of the results. In the area of multilevel thresholding, many types of images are addressed. However, there are a bunch of datasets that are used as a baseline evaluation tool. In this book, we would refer to the following images as our benchmark subset, which contains five images with different types of histograms to analyze the properties of each objective

Table 2.5. Examples of new implementations of energy curve thresholding.

Article	Title	Metaheuristic	Fitness function	Metrics	Type of images
(Patra et al., 2014)	A novel context sensitive multilevel thresholding for image segmentation	Genetic Algorithm (GA)	Entropy criterion	Davies Bouldin index	General images
(Pare et al., 2017)	A multilevel color image segmentation technique based on Cuckoo Search Algorithm and energy curve	Cuckoo Search Algorithm	Kapur entropy	MSE, PSNR, FSIM, SSIM	Color images
(Oliva et al., 2018)	Context based image segmentation using Antlion optimization and sine cosine algorithm	Antlion optimization and sine cosine algorithm	Otsu, Kapur	PSRN, FSIM, SSIM	General images
(Díaz-Cortés et al., 2018)	A multilevel thresholding method for breast thermograms analysis using Dragonfly Algorithm	Dragonfly Algorithm	Otsu, Kapur	PSNR, FSMI, SSIM	Breast thermograms
(Srikanth and Bikshalu, 2021)	Multilevel thresholding image segmentation-based on energy curve with Harmony Search Algorithm	Harmony Search Algorithm	Harmony Search Algorithm	Dunn's indices, Davies-Bouldin Index, SD validity index, probabilistic Rand index, PSNR	

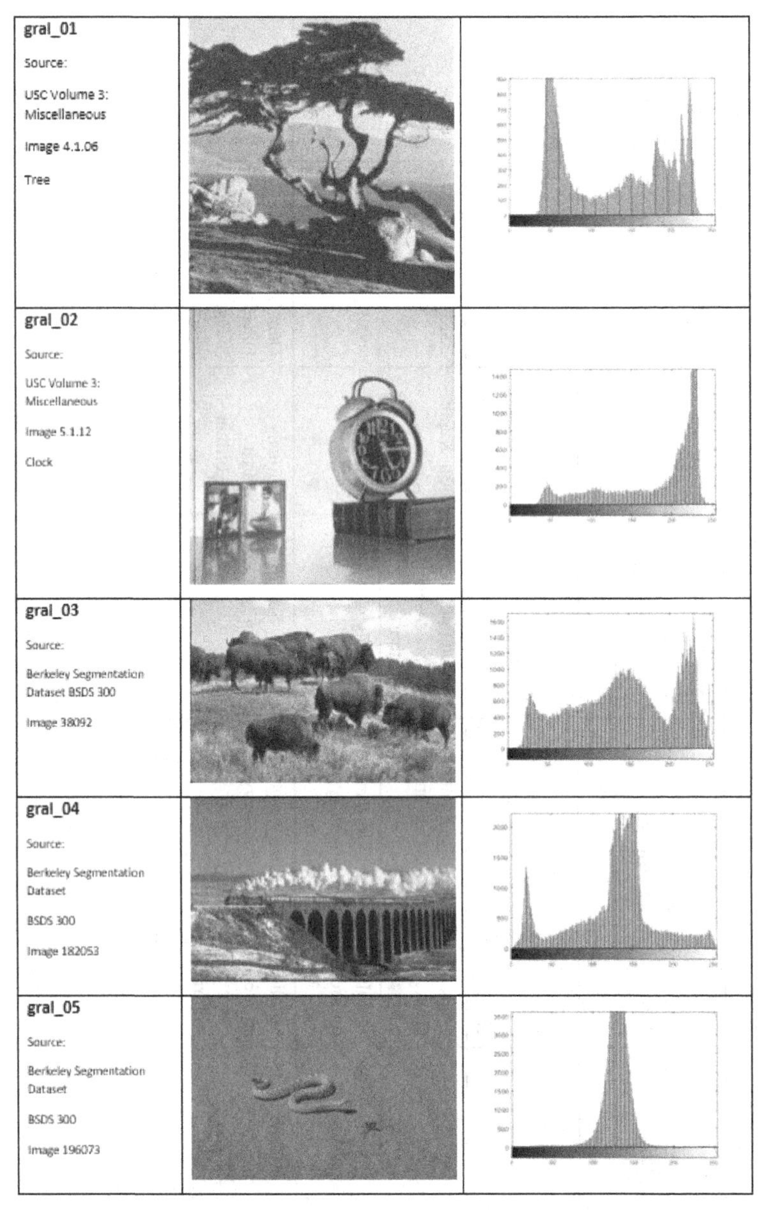

Fig. 2.2. Benchmark images and their histograms.

function and algorithm. The selected images include challenging multimodal histograms except for the image gral_05, where the histogram follows a normal distribution. See Figure 2.2.

2.5 Conclusions

In this chapter we have reviewed some of the most relevant articles over the last five years. The list is by no means extensive but is designed to illustrate the complexity of the area and to highlight future research directions where the reader might contribute. Especially in hyper-heuristic implementations, considering multi-objective approaches, or even new entropy formulations such as the Masi entropy.

Exercises

2.1 Use Google Scholar to find articles about multilevel thresholding over the last year.

2.2 Use Scopus Search to find the most recent metaheuristic algorithms.

2.3 From the results of Exercise 2.1, choose an article and determine in which category would it fall in section 2.3.

2.4 Following the previous exercise, identify which images are used as a benchmark, and search for them over the web.

References

Abd Elaziz, M., Bhattacharyya, S. and Lu, S. (2019). Swarm selection method for multilevel thresholding image segmentation. *Expert Systems with Applications*, 138: 112818. https://doi.org/10.1016/J.ESWA.2019.07.035.

Abd Elaziz, M., Nabil, N., Moghdani, R., Ewees, A.A., Cuevas, E. and Lu, S. (2021). Multilevel thresholding image segmentation based on improved volleyball premier league algorithm using whale optimization algorithm. *Multimedia Tools and Applications*, 80(8): 12435–12468. https://doi.org/10.1007/S11042-020-10313-W/FIGURES/14.

Agrawal, S., Panda, R. and Abraham, A. (2020). A novel diagonal class entropy-based multilevel image thresholding using coral reef optimization. *IEEE Transactions on Systems, Man, and Cybernetics: Systems*, 50(11): 4688–4696. https://doi.org/10.1109/TSMC.2018.2859429.

Ben Ishak, A. (2017). A two-dimensional multilevel thresholding method for image segmentation. *Applied Soft Computing*, 52: 306–322. https://doi.org/10.1016/J.ASOC.2016.10.034.

Bhandari, A.K. and Kumar, I.V. (2019). A context sensitive energy thresholding-based 3D Otsu function for image segmentation using human learning optimization. *Applied Soft Computing*, 82: 105570. https://doi.org/10.1016/J.ASOC.2019.105570.

Borjigin, S. and Sahoo, P.K. (2019). Color image segmentation based on multilevel Tsallis–Havrda–Charvát entropy and 2D histogram using PSO algorithms. *Pattern Recognition*, 92: 107–118. https://doi.org/10.1016/J.PATCOG.2019.03.011.

Díaz-Cortés, M.-A., Ortega-Sánchez, N., Hinojosa, S., Oliva, D., Cuevas, E., Rojas, R. and Demin, A. (2018). A multilevel thresholding method for breast thermograms analysis using Dragonfly Algorithm. *Infrared Physics & Technology*, 93: 346–361. https://doi.org/10.1016/j.infrared.2018.08.007.

Elaziz, M.A., Ewees, A.A. and Hassanien, A.E. (2017). Whale optimization algorithm and moth-flame optimization for multilevel thresholding image segmentation. *Expert Systems with Applications*, 83: 242–256. https://doi.org/10.1016/J.ESWA.2017.04.023.

Elaziz, M.A. and Lu, S. (2019). Many-objectives multilevel thresholding image segmentation using Knee Evolutionary Algorithm. *Expert Systems with Applications*, 125: 305–316. https://doi.org/10.1016/J.ESWA.2019.01.075.

Elaziz, M.A., Oliva, D., Ewees, A.A. and Xiong, S. (2019). Multilevel thresholding-based greyscale image segmentation using multi-objective multiverse optimizer. *Expert Systems with Applications*, 125: 112–129. https://doi.org/10.1016/J.ESWA.2019.01.047.

Elaziz, M.A., Ewees, A.A. and Oliva, D. (2020). Hyperheuristic method for multilevel thresholding image segmentation. *Expert Systems with Applications*, 146: 113201. https://doi.org/10.1016/J.ESWA.2020.113201.

He, L. and Huang, S. (2017). Modified firefly algorithm based multilevel thresholding for color image segmentation. *Neurocomputing*, 240: 152–174. https://doi.org/10.1016/J.NEUCOM.2017.02.040.

Hinojosa, S., Avalos, O., Galvez, J., Oliva, D., Cuevas, E. and Perez-Cisneros, M. (2019). Remote sensing imagery segmentation based on multi-objective optimization algorithms. *2018 IEEE Latin American Conference on Computational Intelligence, LA-CCI 2018*. https://doi.org/10.1109/LA-CCI.2018.8625215.

Hinojosa, S., Avalos, O., Oliva, D., Cuevas, E., Pajares, G., Zaldivar, D. and Gálvez, J. (2018). Unassisted thresholding based on multi-objective evolutionary algorithms. *Knowledge-Based Systems*, 159: 221–232. https://doi.org/10.1016/J.KNOSYS.2018.06.028.

Hinojosa, S., Oliva, D., Cuevas, E., Pajares, G., Zaldivar, D. and Pérez-Cisneros, M. (2020). Reducing overlapped pixels: A multi-objective color thresholding approach. *Soft Computing*, 24(9): 6787–6807. https://doi.org/10.1007/S00500-019-04315-6/FIGURES/7.

Houssein, E.H., Helmy, B.E. din, Oliva, D., Elngar, A.A. and Shaban, H. (2021). A novel Black Widow Optimization algorithm for multilevel thresholding image segmentation. *Expert Systems with Applications*, 167: 114159. https://doi.org/10.1016/J.ESWA.2020.114159.

Houssein, E.H., Hussain, K., Abualigah, L., Elaziz, M.A., Alomoush, W., Dhiman, G., Djenouri, Y. and Cuevas, E. (2021). An improved opposition-based marine predators' algorithm for global optimization and multilevel thresholding image segmentation. *Knowledge-Based Systems*, 229: 107348. https://doi.org/10.1016/J.KNOSYS.2021.107348.

Kapur, J.N.N., Sahoo, P.K.K. and Wong, A.K.C.K.C. (1985). A new method for gray-level picture thresholding using the entropy of the histogram. *Computer Vision, Graphics, and Image Processing*, 29(3): 273–285. https://doi.org/10.1016/0734-189X(85)90125-2.

Karakoyun, M., Gülcü, Ş. and Kodaz, H. (2021). D-MOSG: Discrete multi-objective shuffled gray wolf optimizer for multilevel image thresholding. *Engineering Science and Technology, an International Journal*, 24(6): 1455–1466. https://doi.org/10.1016/J.JESTCH.2021.03.011.

Khairuzzaman, A.K.M. and Chaudhury, S. (2017). Multilevel thresholding using grey wolf optimizer for image segmentation. *Expert Systems with Applications*, 86: 64–76. https://doi.org/10.1016/J.ESWA.2017.04.029.

Khairuzzaman, A.K.M. and Chaudhury, S. (2019). Masi entropy-based multilevel thresholding for image segmentation. *Multimedia Tools and Applications*, 78(23): 33573–33591. https://doi.org/10.1007/S11042-019-08117-8/TABLES/4.

Kumar, M., Bhandari, A.K., Singh, N. and Ghosh, A. (2021). A new multilevel histogram thresholding approach using variational mode decomposition. *Multimedia Tools and Applications*, 80(7): 11331–11363. https://doi.org/10.1007/S11042-020-10189-W/FIGURES/9.

Li, C.H. and Lee, C.K. (1993). Minimum cross-entropy thresholding. *Pattern Recognition*, 26(4): 617–625. https://doi.org/10.1016/0031-3203(93)90115-D.

Li, Y., Fan, X. and Li, G. (2007). Image segmentation based on Tsallis-entropy and Rényi-entropy and their comparison. *2006 IEEE International Conference on Industrial Informatics, INDIN'06*, 00(i): 943–948. https://doi.org/10.1109/INDIN.2006.275704.

Liang, H., Jia, H., Xing, Z., Ma, J. and Peng, X. (2019). Modified Grasshopper algorithm-based multilevel thresholding for color image segmentation. *IEEE Access*, 7: 11258–11295. https://doi.org/10.1109/ACCESS.2019.2891673.

Mittal, H. and Saraswat, M. (2018). An optimum multi-level image thresholding segmentation using non-local means 2D histogram and exponential Kbest gravitational search algorithm. *Engineering Applications of Artificial Intelligence*, 71: 226–235. https://doi.org/10.1016/J.ENGAPPAI.2018.03.001.

Oliva, D., Hinojosa, S., Elaziz, M.A. and Ortega-Sánchez, N. (2018). Context based image segmentation using antlion optimization and sine cosine algorithm. *Multimedia Tools and Applications*, 77(19): 25761–25797. https://doi.org/10.1007/s11042-018-5815-x.

Otsu, N. (1979). A threshold selection method from gray-level histograms. *IEEE Transactions on Systems, Man, and Cybernetics*, 9(1): 62–66. https://doi.org/10.1109/TSMC.1979.4310076.

Pare, S., Kumar, A., Bajaj, V. and Singh, G.K., Pare, S., Kumar, A., Bajaj, V. and Singh, G.K. (2017). An efficient method for multilevel color image thresholding using Cuckoo Search algorithm based on minimum cross-entropy. *Applied Soft Computing*, 61: 570–592. https://doi.org/10.1016/J.ASOC.2017.08.039.

Patra, S., Gautam, R. and Singla, A. (2014). A novel context sensitive multilevel thresholding for image segmentation. *Applied Soft Computing Journal*, 23: 122–127. https://doi.org/10.1016/j.asoc.2014.06.016.

Sahoo, P.K. and Arora, G. (2004). A thresholding method based on two-dimensional Rényi's entropy. *Pattern Recognition*, 37(6): 1149–1161. https://doi.org/10.1016/j.patcog.2003.10.008.

Sarkar, S., Das, S. and Chaudhuri, S.S. (2017). Multilevel thresholding with a decomposition-based multi-objective evolutionary algorithm for segmenting natural and medical images. *Applied Soft Computing*, 50: 142–157. https://doi.org/10.1016/J.ASOC.2016.10.032.

Srikanth, R. and Bikshalu, K. (2021). Multilevel thresholding image segmentation based on energy curve with Harmony Search Algorithm. *Ain Shams Engineering Journal*, 12(1): 1–20. https://doi.org/10.1016/J.ASEJ.2020.09.003.

Tan, Z. and Zhang, D. (2020). A fuzzy adaptive gravitational search algorithm for two-dimensional multilevel thresholding image segmentation. *Journal of Ambient Intelligence and Humanized Computing*, 11(11): 4983–4994. https://doi.org/10.1007/S12652-020-01777-7/TABLES/6.

Vig, G. and Kumar, S. (2021). Comparison of different metaheuristic algorithms for multilevel non-local means 2D histogram thresholding segmentation. *Advances in Intelligent Systems and Computing*, 1086: 563–572. https://doi.org/10.1007/978-981-15-1275-9_46.

Wunnava, A., Naik, M.K., Panda, R., Jena, B. and Abraham, A. (2020a). A novel interdependence based multilevel thresholding technique using Adaptive Equilibrium Optimizer. *Engineering Applications of Artificial Intelligence*, 94: 103836. https://doi.org/10.1016/J.ENGAPPAI.2020.103836.

Wunnava, A., Naik, M.K., Panda, R., Jena, B. and Abraham, A. (2020b). An adaptive Harris Hawks optimization technique for two-dimensional grey gradient based multilevel image thresholding. *Applied Soft Computing*, 95: 106526. https://doi.org/10.1016/J.ASOC.2020.106526.

Xing, Z. and He, Y. (2021). Many-objective multilevel thresholding image segmentation for infrared images of power equipment with boost Marine Predators' algorithm. *Applied Soft Computing*, 113: 107905. https://doi.org/10.1016/J.ASOC.2021.107905.

Yan, Z., Zhang, J. and Tang, J. (2020). Modified water wave optimization algorithm for underwater multilevel thresholding image segmentation. *Multimedia Tools and Applications*, 79(43–44): 32415–32448. https://doi.org/10.1007/S11042-020-09664-1/FIGURES/14.

Yan, Z., Zhang, J., Yang, Z. and Tang, J. (2021). Kapur's entropy for underwater multilevel thresholding image segmentation based on Whale Optimization Algorithm. *IEEE Access*, 9: 41294–41319. https://doi.org/10.1109/ACCESS.2020.3005452.

Zhou, Y., Yang, X., Ling, Y. and Zhang, J. (2018). Metaheuristic Moth Swarm Algorithm for multilevel thresholding image segmentation. *Multimedia Tools and Applications*, 77(18): 23699–23727. https://doi.org/10.1007/S11042-018-5637-X/FIGURES/10.

CHAPTER 3
The Political Optimizer for Image Thresholding

3.1 Introduction

Metaheuristic algorithms have different sources of inspiration, a growing number of proposals are based in human interactions and social processes of human beings. The social algorithms are becoming popular since different methods have proven their performance in complex problems. The Political Optimizer (PO) is a recently proposed metaheuristic that imitates a political system of a country (Askari et al., 2020a). Politics is a complicated process in which are involved different individuals and organizations that are looking for personal and group goals. For example, the candidates are trying to win the election and the parties are looking for having more representatives in the parliament. Besides, each country could have one, two or multiple parties. This also creates different kinds of political systems as single-party, bi-party, multi-party, and dominant-party.

In the PO, the authors propose a generalization of a political system in which six phases are integrated (see Figure 3.1) Such phases are (1) Party formation, (2) Party switching, (3) Ticket allocation, (4) Election campaign, (5) Inter-party election, and (6) Government formation and parliamentary affairs. Besides, in the PO is imitated a multi-party democracy. Since its publication the PO has been applied to solve complex problems in different fields. For example, in energy systems the PO has been used to estimate the parameters of fuel cells (Zaki Diab et al., 2020), for the optimization of the cost of hybrid wind-solar-diesel-battery systems (Singh et al., 2020), and for the estimation of parameters in solar cells and photovoltaic modules (Yousri et al., 2020). Besides it has also been used for the optimization of truss structures (Awad, 2021), and in machine learning (Askari and Younas, 2021; Manita and Korbaa, 2020).

On the other hand, Otsu's between-class variance is a popular method used for image thresholding (Otsu, 1979). This is a non-parametric thresholding approach that considers the maximum value of the variance that exists between the classes; to perform this task it is necessary to obtain the previous histogram of the image. In general terms, the simplest version of this technique performs an exhaustive search for a single threshold value. In this case, the search mechanism is moving the threshold along the histogram values and computing the variance that exist between the classes created at each time. Finally, the maximum value is selected as the best threshold. In the related literature, Otsu's methodology has been extensively used and modified (Khandare and Isalkar, 2014). One of the most extended modifications of Otsu's technique between class variance is its use as an objective function; here different optimization methods have been used (Oliva et al., 2019). The use of Otsu's method in combination with the optimization algorithm has become very popular in recent years. The reason is that by using thresholding for image segmentation it is possible to obtain a fast extraction of the elements contained in the images. For example, it has been proposed an improved invasive weed algorithm for image thresholding by using Otsu's formulation for between-class variance as an objective function. Besides, the use of Otsu's method has been extended for the segmentation of thermal images for breast cancer detection (Díaz-Cortés et al.,

2018). In the same way, the Otsu technique between class variance has been used for the segmentation of satellite images by using a dynamic version of the Harris Hawks optimization algorithm (Jia et al., 2019).

This chapter aims to verify the efficiency of the PO for solving the image thresholding problem by combining it with the Otsu's objective functions. The idea is to check if the algorithm can find the best configuration of thresholds that maximizes the between-class variance for the segmentation of different images. For comparative purposes, they have been used in different metrics that permit a validation of the quality of the segmented images. The images used present different degrees of complexity since the histograms have different distributions.

3.2 The Political Optimizer

Politics is a complicated process since it involves different elements that are looking for independent and collective goals. Besides the interaction of humans and the, behavior is hard do model. The PO simplifies the politics process by considering four aspects that are depicted in Figure 3.1 and listed as follows:

- The electoral process.
- The collaboration and competitiveness inside the party.
- The improvement of candidates based on previous elections.
- The cooperative behavior and interaction of the winners to run a government.

The PO begins by creating a party and the allocation of the constituency. Here is generated a population XP that is divided into N political parties P (see Eq. 3.1). Moreover, each party has n members **S** that are considered solutions (s) with d dimensions (Eqs. 3.2 and 3.3).

$$XP = \left[P_1, P_2, P_3, \ldots, P_N\right] \tag{3.1}$$

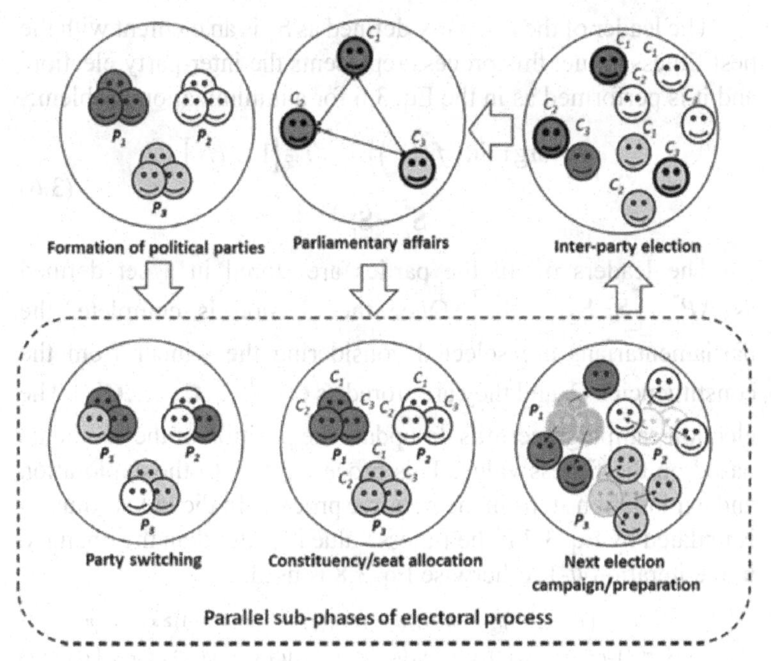

Fig. 3.1. Example of the phases of the political process modeled in the PO (Askari et al., 2020b).

$$P_i = \left[\mathbf{S}_i^1, \mathbf{S}_i^2, \mathbf{S}_i^3, \ldots, \mathbf{S}_i^n \right] \quad (3.2)$$

$$\mathbf{S}_i^j = \left[s_{i,1}^j, s_{i,2}^j, s_{i,3}^j, \ldots, s_{i,d}^j \right] \quad (3.3)$$

Once the parities are initialized, N constituencies are built by using Eq. 3.4.

$$\mathbf{C} = \left[C_1, C_2, C_3, \ldots, C_N \right] \quad (3.4)$$

Each constituency C_j contains a corresponding j-th member of each party as is described in Eq. 3.5.

$$\mathbf{C} = \left[\mathbf{S}_1^j, \mathbf{S}_2^j, \mathbf{S}_3^j, \ldots, \mathbf{S}_N^j \right] \quad (3.5)$$

The leader of the i-th party defined as $\mathbf{S}_i^*$ is an element with the best fitness value; this process represents the inter-party election, and it is performed as in the Eq. 3.6 for a minimization problem:

$$q = \arg\min_{1 \le j \le N} \left(f\left(\mathbf{S}_i^j\right) \right), \quad \forall i \in [1,\ldots,N]$$
$$\mathbf{S}_i^* = \mathbf{S}_i^q \tag{3.6}$$

The leaders of all the parties are stored in a set defined as $XP^* = \left[\mathbf{S}_1^*, \mathbf{S}_2^*, \ldots, \mathbf{S}_N^*\right]$. Once the election is complete, the parliamentarians are selected considering the winner from the constituencies **C**, and they are stored as $\mathbf{C}^* = \left[\mathbf{C}_1^*, \mathbf{C}_2^*, \ldots, \mathbf{C}_N^*\right]$. The election campaign permits to update the position of the candidate based on the fitness value. This process refers to the exploration and exploitation steps in the iterative process. In the PO, a solution is updated by Eq. 3.7 if the fitness value is better than the obtained in the iteration it-1, otherwise Eq. 3.8 is used.

$$s_{i,k}^j(it+1) = \begin{cases} m^* + r\left(m^* - s_{i,k}^j(it)\right), & \text{if } s_{i,k}^j(it-1) \le s_{i,k}^j(it) \le m^* \text{ or } s_{i,k}^j(it-1) \ge s_{i,k}^j(it) \ge m^* \\ m^* + (2r-1)\left|m^* - s_{i,k}^j(it)\right|, & \text{if } s_{i,k}^j(it-1) \le m^* \le s_{i,k}^j(it) \text{ or } s_{i,k}^j(it-1) \ge m^* \ge s_{i,k}^j(it) \\ m^* + (2r-1)\left|m^* - s_{i,k}^j(it-1)\right|, & \text{if } m^* \le s_{i,k}^j(it-1) \le s_{i,k}^j(it) \text{ or } m^* \ge s_{i,k}^j(it-1) \ge s_{i,k}^j(it) \end{cases} \tag{3.7}$$

$$s_{i,k}^j(it+1) = \begin{cases} m^* + (2r-1)\left|m^* - s_{i,k}^j(it)\right|, & \text{if } s_{i,k}^j(it-1) \le s_{i,k}^j(it) \le m^* \text{ or } s_{i,k}^j(it-1) \ge s_{i,k}^j(it) \ge m^* \\ s_{i,k}^j(it-1) + r\left(s_{i,k}^j(it) - s_{i,k}^j(it-1)\right), & \text{if } s_{i,k}^j(it-1) \le m^* \le s_{i,k}^j(it) \text{ or } s_{i,k}^j(it-1) \ge m^* \ge s_{i,k}^j(it) \\ m^* + (2r-1)\left|m^* - s_{i,k}^j(it-1)\right|, & \text{if } m^* \le s_{i,k}^j(it-1) \le s_{i,k}^j(it) \text{ or } m^* \ge s_{i,k}^j(it-1) \ge s_{i,k}^j(it) \end{cases} \tag{3.8}$$

In politics, it is common to change from one party to another. This is also modeled in the PO as a process called party switching. In it, a member $\mathbf{S}_i^j$ that has a probability λ switches to another party P_r that is randomly selected. $\mathbf{S}_i^q$ also swaps its party with the worst element of P_r defined as $\mathbf{S}_r^q$. q is the index of the worst solution and for a minimization problem it is computed as shown in Eq. 3.9:

$$q = \arg\max_{1 \le j \le N} \left(f\left(\mathbf{S}_r^j\right) \right) \tag{3.9}$$

The election is the next step. Here is evaluated the fitness of each contestant of a constituency and it selects the best of them.

This process is performed by using Eq. 3.10 for a minimization problem.

$$q = \arg\min_{1 \le j \le N} \left(f\left(\mathbf{S}_i^j \right) \right), \quad \mathbf{C}_j^* = \mathbf{S}_q^j \tag{3.10}$$

From Eq. 3.10, $\mathbf{C}_j^*$ corresponds to the winner of the constituency $\mathbf{C}_j$. The party leaders then are also updated using Eq. 3.6. Finally, the last process in the PO corresponds to the parliamentary affairs, here the parliamentarian element $\mathbf{C}_j^*$ is updated as follows (Eq. 3.11):

$$\mathbf{C}_{new}^* = \mathbf{C}_r^* + (2a - 1)\left| \mathbf{C}_r^* - \mathbf{C}_j^* \right| \tag{3.11}$$

where r is a random integer in the range 1 to N, $r \neq j$ and a is a random number uniformly distributed between 0 and 1. For the PO, it is important to mention that it was originally proposed for minimization problems, but it can be easily modified for maximization.

3.3 Otsu's Methodology between Class Variance

The Otsu's formulation of between-class variance is a nonparametric segmentation technique proposed in 1979 (Otsu, 1979). In this method it is necessary to find the maximum variance that exists in the classes created by using a threshold. The threshold value helps to divide the pixels of a digital image by using the histogram. Considering a grayscale image with L intensity levels, the probability distribution of the intensity values is obtained as given in Eq. 3.12:

$$Ph_i = \frac{h_i}{NP}, \quad \sum_{i=1}^{NP} Ph_i = 1 \tag{3.12}$$

where i is a specific intensity level ($0 \le i \le L - 1$), NP is the total number of pixels in the image. h_i (represents the histogram) is the number of pixels that correspond to the intensity level of i. The histogram is normalized in a probability distribution Ph_i. For the simplest segmentation (bilevel) two classes are defined as in Eq. 3.13:

$$Cl_1 = \frac{Ph_1}{\omega_0(th)}, \ldots, \frac{Ph_{th}}{\omega_0(th)} \quad \text{and} \quad Cl_2 = \frac{Ph_{th+1}}{\omega_1(th)}, \ldots, \frac{Ph_L}{\omega_1(th)} \tag{3.13}$$

From Eq. 3.13 $\omega_0(th)$ and $\omega_1(th)$ are distributions of probabilities for Cl_1 and Cl_2; such probabilities are computed as in Eq. 3.14.

$$\omega_0(th) = \sum_{i=1}^{th} Ph_i, \quad \omega_1(th) = \sum_{i=th+1}^{L} Ph_i \tag{3.14}$$

To obtain the variance of each class, it is necessary to calculate the mean levels for each class μ_0 and μ_1 by using the next Eq. (3.15):

$$\mu_0 = \sum_{i=1}^{th} \frac{iPh_i}{\omega_0(th)}, \quad \mu_1 = \sum_{i=th+1}^{L} \frac{iPh_i}{\omega_1(th)} \tag{3.15}$$

Now the variances σ_1 and σ_2 of each class are obtained with Eq. 3.16. Besides, the total between class variance σ_{B}^2 is computed using Eq. 3.17.

$$\sigma_1 = \omega_0\left(\mu_0 + \mu_T\right)^2, \quad \sigma_2 = \omega_1\left(\mu_1 + \mu_T\right)^2 \tag{3.16}$$

$$\sigma_{\mathrm{B}}^2 = \sigma_1 + \sigma_2 \tag{3.17}$$

The optimization problem for two classes requires a search mechanism for one *th* value and it is formulated as follows (Eq. 3.18):

$$J(th) = \max(\sigma_B^2(th)), \quad 0 \le th \le L-1 \tag{3.18}$$

By using the previous equations, it is possible to segment an image in two classes according to the pixel's distribution. However, in most of the cases, it is necessary to generate *k* classes; under such circumstances it is defined as a set of thresholds $\mathbf{TH} = [th_1, th_2, \ldots, th_{k-1}]$. This is called multilevel thresholding and under the Otsu's scheme the objective function from Eq. 3.18 is then reformulated as in Eq. 3.19.

$$J(\mathbf{TH}) = \max(\sigma_B^2(\mathbf{TH})), \quad 0 \le th_i \le L-1, \quad i = 1,2,...,k \tag{3.19}$$

The total variance is now computed as follows (Eq. 3.20):

$$\sigma_B^2 = \sum_{i=1}^{k} \sigma_i = \sum_{i=1}^{k} \omega_i \left(\mu_i - \mu_T\right)^2 \tag{3.20}$$

Notice that for Eq. 3.20, i is the index for a specific class. The probability distributions ω_i are computed separately for each class by using Eq. 3.21.

$$\omega_0(th) = \sum_{i=1}^{th_1} Ph_i, \quad \omega_1(th) = \sum_{i=th_1+1}^{th_2} Ph_i, \quad ..., \quad \omega_{k-1}(th) = \sum_{i=th_k+1}^{L} Ph_i \tag{3.21}$$

For the mean values μ_j^c, a similar procedure is used as described in Eq. 3.22.

$$\mu_0 = \sum_{i=1}^{th_1} \frac{iPh_i}{\omega_0(th_1)}, \quad \mu_1 = \sum_{i=th_1+1}^{th_2} \frac{iPh_i}{\omega_1(th_1)}, \quad ..., \quad \mu_{k-1} = \sum_{i=th_k+1}^{L} \frac{iPh_i}{\omega_{k-1}(th_k)} \tag{3.22}$$

3.4 Multilevel Thresholding with PO and Otsu

This section introduces the implementation of the PO for multilevel image thresholding by using the Otsu's between-class variance. The solutions are containing decision variables that correspond to a set of thresholds. The population XP in the PO is defined as in Eq. 3.23:

$$XP = \left[\mathbf{TH}_1, \mathbf{TH}_2, \mathbf{TH}_3, ..., \mathbf{TH}_N\right], \quad \mathbf{TH}_i = P_i = \left[th_1, th_2, ..., th_k\right]^T \tag{3.23}$$

Figure 3.2 shows the flowchart of the PO implementation for digital image thresholding by using the between-class variance of Otsu's methodology. The stop criterion in Figure 3.2 is the maximum number of iterations (*Maxiter*). When *It* reaches the maximum value, the algorithm ends and the best set of thresholds is extracted and applied to the orifical image in grayscale. Notice, that this procedure could be extended for color images as RGB (red, green, blue) by applying the method over each channel separately.

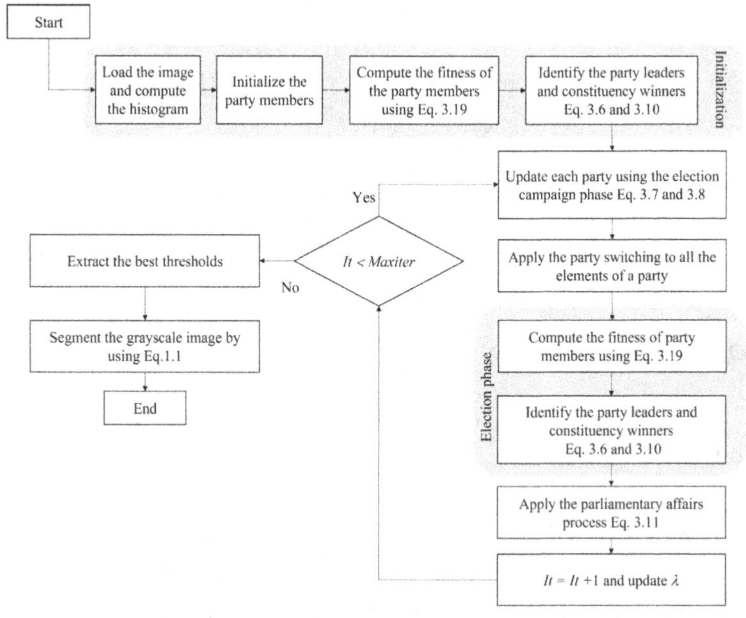

Fig. 3.2. Flowchart of the PO combined with Otsu's between-class variance for digital image thresholding.

3.5 Experiments

The implementation of the PO for image thresholding previously described has been tested over the set of benchmark images described in Chapter 2. The PO was configured in the standard version suggested by the authors in (Askari et al., 2020b). For thresholding, they have been selected four different values th = 2, 4, 8, 16. Such amounts of thresholds help to verify the performance of the PO in multidimensional problems. Since the PO has stochastic variables a set of 35 independent experiments were performed over each image by using every threshold value. Besides, the segmented images are also analyzed by using the quality metrics, Peak Signal to Noise Ratio (PSNR) (Sankur et al., 2002), the Structural Similarity Index (SSIM), and the

Feature Similarity Index (FSIM) that were described in Table 1.1 (Chapter 1). The segmented images are presented in Figure 3.3.

From Figure 3.3 it is possible to see the different shapes and complexities of the histograms. They are completely different from one image to other. However, in all the cases the PO was able to find the optimal configuration of thresholds. It is important to mention that the influence of the number of thresholds in the segmentation permits have better outputs because the classification of the pixels is more detailed. A good comparison could be done between the output images of $th = 4$ and $th = 8$, here the images segmented by 8 thresholds are visually better. Besides, we could also see how the thresholds start distributing in low levels of segmentation according to the histogram. Some good examples are the histograms of images gral_03 and gral_04, they have different complexities and with $th = 2$ the classes are clearer to see. But when $th = 16$ it is more difficult to visually see the classes. However, since the histograms are not completely flat lines, the spikes that they possess are also classified by the thresholds when their number increases. Regarding the quality of the outputs, Table 3.1 presents the results generated by the output images in terms of the PSNR, SSIM, and FSIM by using the PO in combination with Otsu's fitness function.

Table 3.1 presents the mean and the standard deviation (std) of the PSNR, SSIM, and FSIM values obtained by comparing the original grayscale image with the segmented image obtained by the thresholds computed using the PO. First, the results in the table are computed from a set of 35 independent experiments. The std permits one to see the stability of the algorithm: in lower dimension as in $th = 2$ the std values are closer to 0 and it means that the PO is more stable with the results. Meanwhile, for $th = 16$, the PO presents some fluctuations that are reflected in std values close to 1. Regarding the mean values of the quality metrics, for all of them when the number of thresholds increases, the values also increase. For PSNR, SSIM, and FSIM it represents a higher

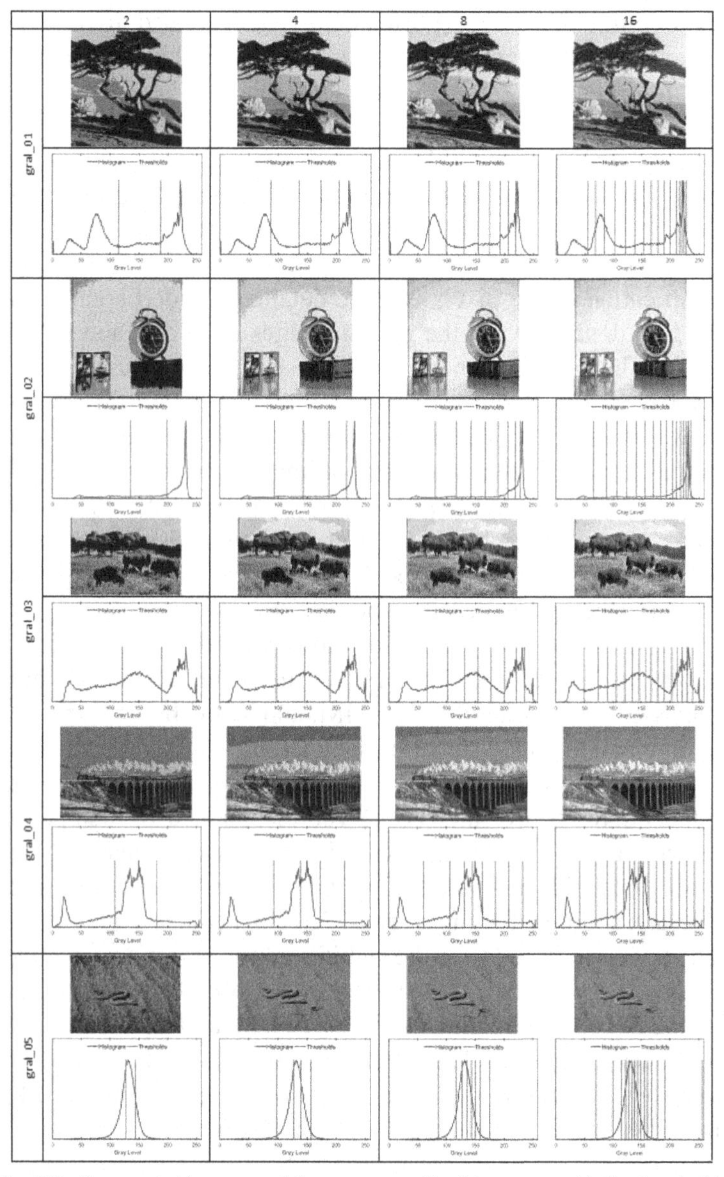

Fig. 3.3. Segmented images and the corresponding histogram with the thresholds obtained by using the PO with the Otsu between-class variance.

Table 3.1. PSNR, SSIM, and FSIM values on general benchmark images using Otsu's between-class variance.

Image	nTh	PO Otsu					
		PSNR		SSIM		FSIM	
		mean	std	mean	std	mean	std
gral_01.tiff	2	13.9663	7.209E-15	0.5400	1.126E-16	0.7401	3.379E-16
	4	16.2878	1.333E-02	0.6397	8.221E-04	0.8312	4.112E-04
	8	18.7053	4.695E-01	0.7401	1.768E-02	0.8692	3.177E-03
	16	22.4665	9.676E-01	0.8462	1.949E-02	0.8842	6.944E-03
gral_02.tiff	2	14.3583	1.262E-14	0.7574	3.379E-16	0.7679	4.506E-16
	4	19.3207	1.807E-02	0.8467	4.146E-04	0.8088	2.052E-04
	8	23.4920	3.291E-01	0.9136	4.533E-03	0.8946	3.389E-03
	16	27.1617	7.498E-01	0.9530	4.330E-03	0.9462	2.001E-03
gral_03.jpg	2	13.4643	7.209E-15	0.5578	4.506E-16	0.6277	2.253E-16
	4	16.7373	4.598E-02	0.6875	1.387E-03	0.7469	9.568E-04
	8	22.1987	2.341E-01	0.8510	5.605E-03	0.8794	4.401E-03
	16	27.1648	5.924E-01	0.9306	7.961E-03	0.9452	5.023E-03
gral_04.jpg	2	14.9481	9.011E-15	0.7248	3.379E-16	0.6763	1.126E-16
	4	17.8164	2.402E-04	0.8021	3.986E-05	0.7875	5.621E-05
	8	23.9649	4.562E-01	0.9053	6.048E-03	0.9075	6.268E-03
	16	28.4746	3.558E-01	0.9437	2.698E-03	0.9603	2.472E-03
gral_05.tiff	2	11.1968	1.802E-15	0.3206	0.000E+00	0.5089	1.126E-16
	4	22.7446	2.151E-01	0.8448	6.084E-03	0.8241	5.895E-03
	8	28.4018	8.668E-01	0.9535	1.009E-02	0.9226	1.276E-02
	16	33.8336	1.225E+00	0.9870	4.212E-03	0.9713	7.722E-03

similarity between the two images. Specifically, for SSIM and FSIM, when its values are closer to 1 the internal structure and features of the pixels from the segmented image are closer to the original image. Besides, it is possible to see that for some images, as the so-called "gral_01.tiff", the values are lower than the in the image called "gral_05.tiff". This situation occurs due to the histogram of the image and the pixel's distribution. In other words, the elements contained in the images "gral_01.tiff" are more difficult to classify that the ones in the scene of "gral_05. tiff". This is easy to visually analyze since "gral_05.tiff" is just a snake in the sand with lower attributes.

3.6 Conclusions

The Otsu's between-class variance is a good segmentation method that could easily find the best thresholds by only using the histogram of a digital image. The popularity of this approach is due to its simplicity; the code is also available in different repositories in the web, and is included in toolboxes and frameworks of image segmentation.

The PO is a recently proposed methodology that is considered as a modern metaheuristic. It is a promising algorithm that still needs to be tested over real problems. In the case of image segmentation, this method is a good alternative since it is easy to adapt with the Otsu function. It is important to mention that each image with a specific threshold value is a different optimization problem by itself. Then The PO can optimize the objective function even in a higher dimension. Another important feature is that the code is free for download. We are sure that in the future it will be adapted to other programming languages.

Exercises

3.1 Implement the Otsu's between-class variance for a single threshold presented in Eq. 3.18 in any programming language.

3.2 Use the exhaustive search to find the best threshold of the cameraman image using the between-class variance for a single threshold.

3.3 Program the Otsu's between-class variance for multilevel thresholding presented in Eq. 3.19.

3.4 Implement the PO in combination with the Otsu's between-class variance for multilevel thresholding and test it over different images.

3.5 Program the PSNR, SSIM, and FSIM and verify the quality of the segmentation.

3.6 Implement the program created in Exercise 3.4 for color images.

References

Askari, Q., Younas, I. and Saeed, M. (2020a). Political Optimizer: A novel socio-inspired metaheuristic for global optimization. *Knowledge-Based Systems*, 195: 105709. https://doi.org/10.1016/j.knosys.2020.105709.

Askari, Q., Younas, I. and Saeed, M. (2020b). Political Optimizer: A novel socio-inspired metaheuristic for global optimization. *Knowledge-Based Systems*, 195: 105709. https://doi.org/10.1016/j.knosys.2020.105709.

Askari, Q. and Younas, I. (2021). Political Optimizer based feedforward neural network for classification and function approximation. *Neural Processing Letters*, 53(1): 429–458. https://doi.org/10.1007/s11063-020-10406-5.

Awad, R. (2021). Sizing optimization of truss structures using the political optimizer (PO) algorithm. *Structures*, 33(June): 4871–4894. https://doi.org/10.1016/j.istruc.2021.07.027.

Díaz-Cortés, M.A., Ortega-Sánchez, N., Hinojosa, S., Oliva, D., Cuevas, E., Rojas, R. and Demin, A. (2018). A multilevel thresholding method for breast thermograms analysis using Dragonfly algorithm. *Infrared Physics and Technology*, 93(May): 346–361. https://doi.org/10.1016/j.infrared.2018.08.007.

Jia, H., Lang, C., Oliva, D., Song, W. and Peng, X. (2019). Dynamic Harris Hawks optimization with mutation mechanism for satellite image segmentation. *Remote Sensing*, 11(12). https://doi.org/10.3390/rs11121421.

Khandare, S.T. and Isalkar, A.D. (2014). A survey paper on image segmentation with thresholding. *International Journal of Computer Science and Mobile Computing*, 3(1): 441–446.

Manita, G. and Korbaa, O. (2020). Binary political optimizer for feature selection using gene expression data. *Computational Intelligence and Neuroscience*, 2020. https://doi.org/10.1155/2020/8896570.

Oliva, D., Elaziz, M.A. and Hinojosa, S. (2019). *Metaheuristic Algorithms for Image Segmentation: Theory and Applications*. Springer.

Otsu, N. (1979). A threshold selection method from gray-level histograms. *IEEE Transactions on Systems, Man, and Cybernetics*, 9(1): 62–66. https://doi.org/10.1109/TSMC.1979.4310076.

Sankur, B., Sankur, B. and Sayood, K. (2002). Statistical evaluation of image quality measures. *Journal of Electronic Imaging*, 11(2): 206. https://doi.org/10.1117/1.1455011.

Singh, P., Pandit, M. and Srivastava, L. (2020). Optimization of levelized cost of hybrid wind-solar-diesel-battery system using political optimizer. *Proceedings of 2020 IEEE 1st International Conference on Smart Technologies for Power, Energy and Control, STPEC 2020*, 1(4): 1–6. https://doi.org/10.1109/stpec49749.2020.9297767.

Yousri, D., Abd Elaziz, M., Oliva, D., Abualigah, L., Al-qaness, M.A.A. and Ewees, A.A. (2020). Reliable applied objective for identifying simple and detailed photovoltaic models using modern metaheuristics: Comparative study. *Energy Conversion and Management*, 223(July): 113279. https://doi.org/10.1016/j.enconman.2020.113279.

Zaki Diab, A.A., Tolba, M.A., Abo El-Magd, A.G., Zaky, M.M. and El-Rifaie, A.M. (2020). Fuel cell parameters estimation via marine predators and political optimizers. *IEEE Access*, 8: 166998–167018. https://doi.org/10.1109/ACCESS.2020.3021754.

Chapter 4
Multilevel Thresholding by Using Manta Ray Foraging Optimization

4.1 Introduction

In the sea there are several species of marine animals, such as sharks, whales, dolphins, manta rays, and other interesting species. All have different behaviors that assists them to survive in the seas, oceans, and large water bodies. Such behaviors help them in the migration process in different seasons or hunt according to their diets. These behaviors are the source of inspiration for researchers in artificial intelligence. Some examples are the whale optimization algorithm (WOA) that mimics the hunting encircling strategy of humpback whales (Mirjalili and Lewis, 2016). The hunting ability of sharks based on their smell sense is also used as a model in the shark smell optimization algorithm (SSO) (Mohammad-Azari et al., 2018). The yellow saddle goatfish (YSG) also has an interesting mechanism to catch its prey; it is used to construct an optimization algorithm called YSGA (Zaldívar et al., 2018). On the other hand, a general behavior of marine predators is formulated in an algorithm called MPA (Faramarzi et al., 2020).

Recently, the manta ray foraging optimization (MRFO) algorithm was proposed as an alternative for solving complex optimization problems (Zhao et al., 2020). The MRFO is based on the different foraging strategies of manta rays. This method is attracting the attention of researchers and it has been widely used and modified for solving problems in different fields of application. Originally the MRFO was introduced for solving engineering benchmark problems (Zhao et al., 2020). In energy systems, it has been employed for the estimation of parameters in solar cells using diode models (Houssein et al., 2021) and as a maximum power point tracker (MPPT) also for solar cells (Fathy et al., 2020). In medicine, it has been applied to classify ECG signals of patients with arrhythmia (Houssein et al., 2021). The MRFO has also been modified for solving different drawbacks as its local stagnation. For example, in (Tang et al., 2021), an improved MRFO is proposed by using a distribution estimation strategy for global optimization problems. Different authors have modified the MRFO for image segmentation for benchmark images and for COVID-19 CT images (Abd Elaziz et al., 2021; Houssein et al., 2021). Such approaches employed the Otsu's between—class variance and the Tsallis fuzzy entropy, respectively.

As was previously mentioned, an easy way to segment an image is by using thresholds over the histogram. Since the histogram could be treated as a probability distribution, different statistical tools can be used to measure the information in the classes. The entropy then is commonly applied for this purpose. The Kapur's entropy is one of the most popular methods used in image thresholding (Kapur et al., 1985). Kapur's idea is to maximize the entropy as a metric of the homogeneity of the classes. Kapur and Otsu methods are probably the most popular techniques used in image thresholding. They share some similarities in terms of the methodology. For example, the simplest perspective Kapur's entropy employs is an exhaustive search to find the optimal thresholds (one or two). When the amount of threshold increases, the computational effort also raises. For that reason, it is necessary to employ a search mechanism that permits to obtain the best threshold configuration. This kind of entropy has been extensively used as objective function and has been combined with different optimization approaches. For example, it has been used with Harmony Search and electromagnetism-like optimization algorithm for the segmentation of benchmark images (Oliva et al., 2014; Oliva et al., 2013). In the same context, it has also been used in the crow search algorithm to explore for the proper configuration of thresholds by maximizing Kapur's entropy (Upadhyay and Chhabra, 2020). This entropy has also been applied for color segmentation by using a hybrid Whale Optimization Algorithm (Lang and Jia, 2019).

This chapter analyzes the performance of the MRFO for image segmentation by using thresholding with Kapur's entropy. It provides an explanation for modifying the MRFO step-by-step to properly find the best thresholds that maximize the objective of Kapur's objective function. Different metrics are used to validate the quality of the segmented images in comparison with the original inputs.

4.2 The Manta Ray Foraging Optimization

The MRFO was introduced as an alternative approach for solving complex problems. The MRFO imitates the behavior of manta rays for catching their preys in the sea. They are three steps in the foraging process called: chain foraging, cyclone foraging, and somersault foraging. Chain foraging permits to search for food sources and when the amount of food is significant the cyclone foraging phase is activated. At this step a spiral movement is created that simulates a cyclone eye that encircles the preys. This permits the manta rays to eat the plankton in circular movements in the somersault foraging process.

The MRFO starts by generating a set of random solutions distributed between the search space. The best element of the population called x_{best} is extracted after the initialization which is based on the fitness valúe. All the phases of the MRFO are explained as follows:

- *Chain foraging*. Here a solution defined as x_k is updated in the iteration k by using the following equation:

$$x_k^{t+1} = \begin{cases} x_k^t + r \cdot \left(x_{best}^t - x_k^t\right) + \alpha \cdot \left(x_{best}^t - x_k^t\right), & k = 1 \\ x_k^t + r \cdot \left(x_{k-1}^t - x_k^t\right) + \alpha \cdot \left(x_{best}^t - x_k^t\right), & k = 2,3,\ldots,N \end{cases} \quad (4.1)$$

$$\alpha = 2 \cdot r \cdot \sqrt{\left|\log(r)\right|}$$

From Eq. 4.1, x_k^t is the *k-th* element at the iteration *t*, *r* is a rando vector with values uniformly distributed between 0 and 1. The variable α is weight coefficient and x_{best}^t corresponds to a region in the search space with higher concentration of plankton.

- *Cyclone foraging*. This process occurs when manta rays identify the plankton; thereafter they form a chain and move spirally through the food source. Here it is important to mention that each manta ray follows the one that is moving in front. In this sense, this is a complex movement around the food, and it is modeled by using the following equation:

$$x_k^{t+1} = \begin{cases} x_{best}^t + r \cdot \left(x_{best}^t - x_k^t\right) + \beta \cdot \left(x_{best}^t - x_k^t\right), & k = 1 \\ x_{best}^t + r \cdot \left(x_{k-1}^t - x_k^t\right) + \beta \cdot \left(x_{best}^t - x_k^t\right), & k = 2,3,\ldots,N \end{cases} \quad (4.2)$$

$$\beta = 2 \cdot \exp\left(r \cdot \frac{K_{\max} - k + 1}{K_{\max}}\right) \cdot \sin(2\pi r)$$

In Eq. 4.2 β, there is also a weight variable. On the other hand, by using a random position in the search space, the cyclone foraging phase permits to obtain more exploration in the search process. This is a desired behavior in metaheuristics because it permits avoidance of the stagnation in suboptimal solutions. Equation 4.2 then could be rewritten as follows (Eq. 4.3):

$$x_k^{t+1} = \begin{cases} x_{rand}^t + r \cdot \left(x_{rand}^t - x_k^t\right) + \beta \cdot \left(x_{rand}^t - x_k^t\right), & k = 1 \\ x_{rand}^t + r \cdot \left(x_{k-1}^t - x_k^t\right) + \beta \cdot \left(x_{rand}^t - x_k^t\right), & k = 2,3,\ldots,N \end{cases} \quad (4.3)$$

$$x_{rand}^t = lb + r \cdot (ub - lb)$$

where lb and ub are the lower and upper bounds of the search space, respectively.

- *Somersault foraging*. In this phase the food source is considered as a fixed point in the search space. Each candidate solution (manta ray) moves around the fixed point and somersaults to a new position in the search space. This movement is mathematically modeled in Eq. 4.4.

$$x_k^{t+1} = x_k^t + s \cdot \left(r1 \cdot x_{best}^t - r2 \cdot x_k^t\right) \quad (4.4)$$

From Eq. 4.4, s is a somersault constant that is set to 2, $r1$ and $r2$ are random variables defined between 0 and 1.

As was previously mentioned, for metaheuristic algorithms, a good balance is expected between the exploration and exploitation. In the MRFO, the switching between exploration and exploitation steps is done by the relationship between current iteration and maximum number iteration (*currentIt/maxit*). The exploitation is then activated when *currentIt/maxit* < *rand*, here the most suitable solution is the fixed position (pivot). On the other hand, the exploration phase is activated *currentIt/maxit* > *rand*. By using the random (*rand*) variable the MRFO can change between the three different foraging phases.

4.3 The Kapur Entropy

In the literature related to image thresholding, it is very common to find different nonparametric techniques based on entropies.

One of the most popular is the entropy proposed by Kapur (Kapur et al., 1985). Akin to the between-class variance of Otsu, the Kapur entropy uses probability distribution of the histogram to compute the thresholds that divide the pixels of the image into different classes. In a simple way, it is considered a grayscale image with L intensity levels. If we want to divide the image into two classes, it is necessary to compute a single threshold (th). Under Kapur's methodology, an exhaustive search is applied in which the entropy is computed for each intensity level of the probability distribution (e.g., from 0 to 256 level). The entropy values are stored in a vector and when all of them are computed the maximum value is selected and the intensity level associated to it corresponds to the th that segments the image (Figure 4.1).

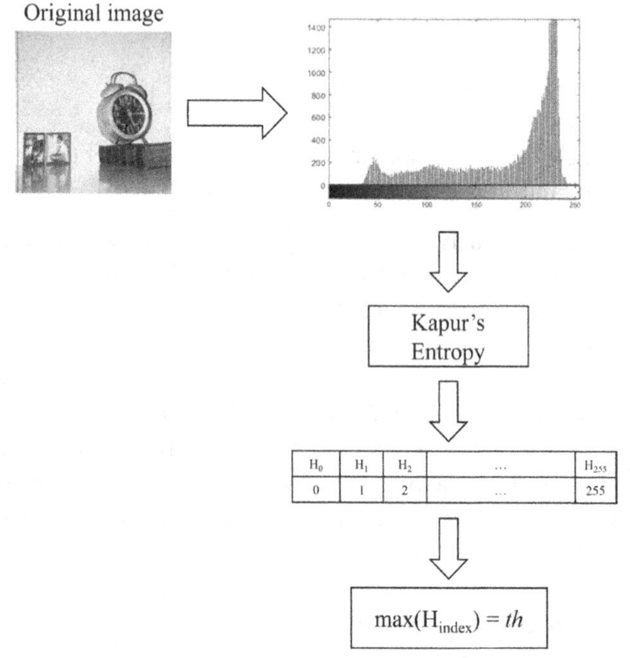

Fig. 4.1. Process of threshold computation by using the histogram and Kapur's entropy.

The exhaustive search is a process that works for one and probably two thresholds, but this is not viable when more classes are required. When more groupings occur, the computational complexity and time drastically increases. To handle these drawbacks, the use of metaheuristics and Kapur's entropy could be integrated and then reformulated as an objective function as in Eq. 4.5 for bi-level thresholding.

$$J(th) = H_1 + H_2 \tag{4.5}$$

In Eq. 4.5 (like Figure 4.1) H_1 and H_2 are the entropies computed as:

$$H_1 = \sum_{i=1}^{th} \frac{Ph_i}{\omega_o} \ln\left(\frac{Ph_i}{\omega_o}\right), \quad H_2 = \sum_{i=th+1}^{L} \frac{Ph_i}{\omega_1} \ln\left(\frac{Ph_i}{\omega_1}\right) \tag{4.6}$$

From Eq. 4.6, Ph_i is defined as the probability distribution of the intensity levels and it is computed by using the following Eq. 4.7:

$$Ph_i = \frac{h_i}{NP}, \quad \sum_{i=1}^{NP} Ph_i = 1 \tag{4.7}$$

where h is the histogram, i is a specific intensity level ($0 \leq i \leq L - 1$), and NP corresponds to the number of pixels. $\omega_0(th)$ and $\omega_1(th)$ are probabilities' distributions for each class and they are computed as in Otsu's between-class variance (Eq. 3.14), ln(.) corresponds to the natural logarithm.

Kapur's entropy can be extended for more than two classes by using multiple thresholds. The objective function can be rewritten as in Eq. 4.8.

$$J(\mathbf{TH}) = \sum_{i=1}^{cl} H_i \tag{4.8}$$

$\mathbf{TH} = [th_1, th_2, \ldots, th_{cl-1}]$, contains the multiple thresholds. Each entropy is computed separately with its respective th value, Eq. 4.6 is updated for cl entropies as follows (Eq. 4.9):

$$H_1 = \sum_{i=1}^{th_1} \frac{Ph_i}{\omega_o} \ln\left(\frac{Ph_i}{\omega_o}\right), \; H_2^c = \sum_{i=th_1+1}^{th_2} \frac{Ph_i}{\omega_1} \ln\left(\frac{Ph_i}{\omega_1}\right), \; \ldots, \; H_k = \sum_{i=th_{cl}+1}^{L} \frac{Ph_i}{\omega_{k-1}} \ln\left(\frac{Ph_i}{\omega_{cl-1}}\right) \tag{4.9}$$

4.4 Multilevel Thresholding with MRFO and Kapur

Here is explained the implementation of MRFO for multilevel thresholding with the use of Kapur's entropy. First, it is necessary to define the codification of the solutions according to the thresholds that in this case are the decision variables. The population and each solution of the MRFO are defined as follows (Eq. 4.10):

$$X = [\mathbf{TH}_1, \mathbf{TH}_2, \mathbf{TH}_3, \ldots, \mathbf{TH}_N], \quad \mathbf{TH}_i = x_k = [th_1, th_2, \ldots, th_k]^T \quad (4.10)$$

The flowchart of the implementation of MRFO with Kapur for image segmentation is presented in Figure 4.2. As in other similar approaches, the stop criterion is the maximum number of iterations. When this value is reached, the MRFO stops the search process and the best configuration of thresholds is extracted and applied over the histogram of the original image.

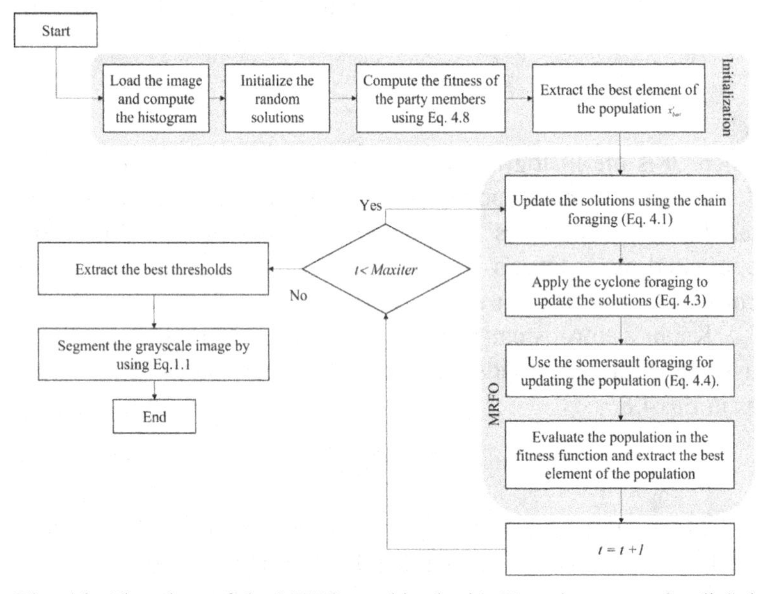

Fig. 4.2. Flowchart of the MRFO combined with Kapur's entropy for digital image thresholding.

4.5 •Experiments

The set of images presented in Chapter 2 is used to test the efficiency of the implementation of the MRFO for digital image thresholding. The MFRO uses the standard configuration suggested in the article where it was first introduced (Zhao et al., 2020). Four amounts of thresholds are defined to test the capabilities of the proposed implementation. In this case th = 2, 4, 8, 16, and they were selected to test the MRFO under different multidimensional circumstances. In segmentation, it is necessary to perform an analysis of the outputs to verify its quality. The Peak Signal to Noise Ratio (PSNR) (Sankur et al., 2002), the Structural Similarity Index (SSIM), and the Feature Similarity Index (FSIM) are used here for the purpose of quality analysis. Such quality metrics have been described in Table 1.1 of Chapter 1. The segmented image and their corresponding histograms with the thresholds obtained by the MRFO are presented in Figure 4.3.

Figure 4.3 shows the histograms of the benchmark images. The histograms present different shapes and complexities. The MRFO has the capabilities to find the optimal set of thresholds for each image in all the experiments. By using different amounts of thresholds, the segmentation is more accurate. In other words, the outputs of th = 2, 4 have less visual quality that the ones obtained by th = 8, 16. This is because with more thresholds it is possible to have more classes and the details of the images are better classified. A good example of a proper segmentation is in image gral_04 with th = 2, here is possible to see the thresholds which are placed permitting generation of the three classes. For the same image using th = 16, the good generation of classes permits to have more details, see for example, the second class from the first to the second threshold. Depending on the histogram, the segmentation could be better or not. However, the MRFO can obtain good segmentation results. Table 4.1 presents the results generated by the output images in terms of the PSNR, SSIM, and FSIM by using the MRFO in combination with Kapur's fitness function.

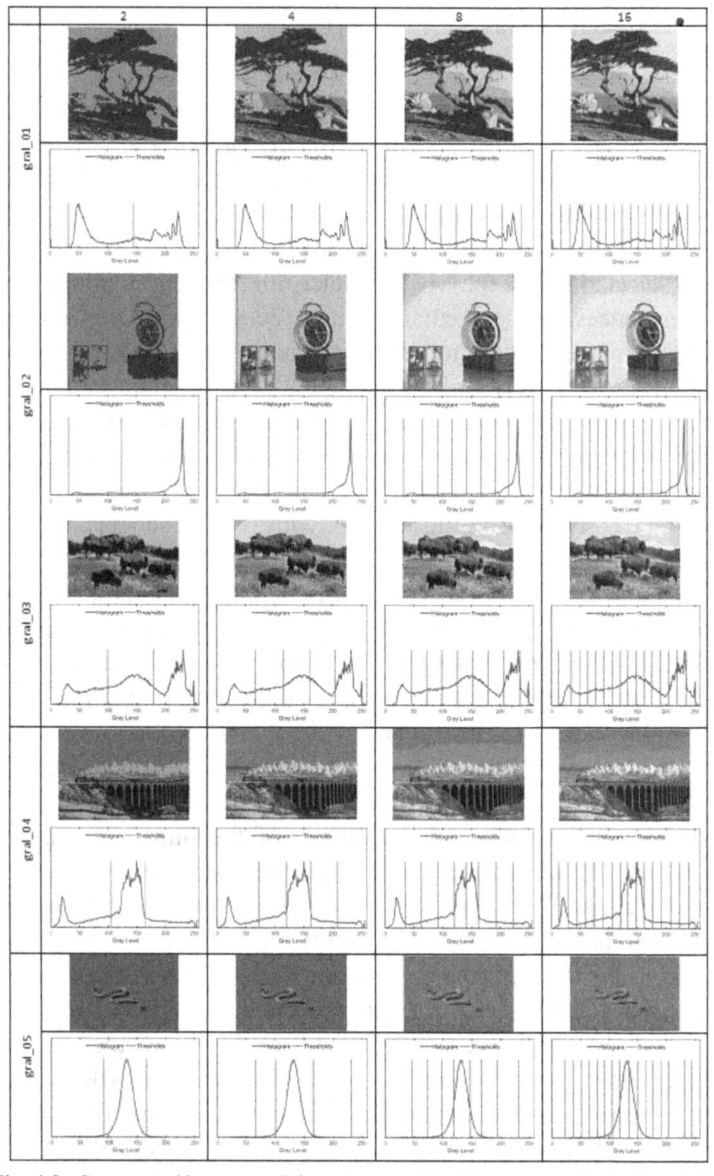

Fig. 4.3. Segmented images and the corresponding histogram with the thresholds obtained by using the MRFO with Kapur's entropy.

Table 4.1. PSNR, SSIM, and FSIM values on general benchmark images using Kapur's entropy.

Image	nTh	PO Otsu					
		PSNR		SSIM		FSIM	
		mean	std	mean	std	mean	std
gral_01.tiff	2	13.7593	1.081E-14	0.7116	0.000E+00	0.6503	2.253E-16
	4	19.0794	7.209E-15	0.8680	5.632E-16	0.8368	5.632E-16
	8	23.2828	3.469E-01	0.9162	4.645E-03	0.9061	6.233E-03
	16	28.8052	9.517E-01	0.9645	7.611E-03	0.9597	6.995E-03
gral_02.tiff	2	9.5547	1.802E-15	0.7541	5.632E-16	0.7307	4.506E-16
	4	17.7908	4.259E-01	0.9148	1.325E-03	0.8353	3.137E-03
	8	25.1152	5.675E-01	0.9558	8.253E-04	0.8977	1.940E-03
	16	29.8480	4.059E-01	0.9757	1.452E-03	0.9457	4.014E-03
gral_03.jpg	2	14.1116	9.011E-15	0.6107	3.379E-16	0.6492	4.506E-16
	4	19.0676	1.081E-14	0.7866	4.506E-16	0.8100	5.632E-16
	8	23.2839	3.561E-02	0.8859	1.091E-03	0.9038	5.981E-04
	16	29.6858	2.410E-01	0.9727	2.540E-03	0.9686	2.304E-03
gral_04.jpg	2	14.9326	1.802E-15	0.7277	2.253E-16	0.6730	3.379E-16
	4	19.3557	0.000E+00	0.8509	4.506E-16	0.8306	3.379E-16
	8	23.8649	1.138E+00	0.9185	4.364E-03	0.9207	1.505E-03
	16	29.6446	7.292E-01	0.9786	1.198E-03	0.9691	2.785E-03
gral_05.tiff	2	15.7291	7.209E-15	0.7009	5.632E-16	0.5126	1.126E-16
	4	18.0582	3.605E-15	0.7538	5.632E-16	0.5883	3.379E-16
	8	24.4588	8.177E-01	0.8957	1.620E-02	0.8991	2.518E-02
	16	30.4622	5.753E-01	0.9677	4.211E-03	0.9777	2.882E-03

The mean and standard deviation (std) of the PSNR, SSIM, and FSIM are presented in Table 4.1. Such values were obtained after performing a comparison between the original image in grayscale and the segmented image obtained after applying the thresholds computed using the MRFO with Kapur's entropy. To obtain the mean and std, 35 independent experiments over each image using the different amounts of thresholds were carried out. The std helps to verify the stability of the results generated using the MRFO. In most of the cases the values are closer to zero, but they are especially lower in th = 2, 4. Regarding the mean values of the quality metrics, for all of them, when the number of thresholds increases, the values also increase. In the case of PSNR, SSIM, and FSIM the higher values are interpreted as a good similarity between the two images. For the SSIM and FSIM, when internal information of the image (features, structures, etc.) is similar between the two images, the values are closer to 1. From Table 4.1, it is possible to see that for some imagessuch as "gral_01.tiff" the values are lower than in the image called "gral_05.tiff". This situation occurs due to the histogram of the image and the pixel's distribution.

4.6 Conclusions

Kapur's entropy is a good choice for image segmentation. Akin to other approaches, it uses the information from the histogram of a digital image. This approach is very popular because it is simple to implement. Nowadays, it is possible to find it in different toolboxes and frameworks of image processing.

On the other hand, the MRFO is a modern metaheuristic algorithm that combines different strategies to solve complex optimization problems. It is a good alternative optimization, and has been used in several real problems. However, its efficiency needs to be tested over different applications. The implementation of MRFO with Kapur's entropy permits verification of the good capabilities of the MRFO. But also, it permits one to see that this is a flexible algorithm that could be easily adapted to different problems. The code of the MRFO is also available online and this is an advantage that empowers it to increase its popularity.

Exercises

4.1 Implement Kapur's entropy for a single threshold presented in Eq. 4.5 in any programming language.

4.2 Use the exhaustive search to find the best threshold of the cameraman image using Kapur's entropy for a single threshold.

4.3 Program the Kapur entropy for multilevel thresholding presented in Eq. 4.8.

4.4 Implement the MRFO in combination with Kapur's entropy for multilevel thresholding and test it over different images.

4.5 Program the PSNR, SSIM, and FSIM and verify the quality of the segmentation.

4.6 Implement the program created in Exercise 4.4 for color images.

References

Abd Elaziz, M., Yousri, D., Al-qaness, M.A.A., AbdelAty, A.M., Radwan, A.G. and Ewees, A.A. (2021). A Grunwald–Letnikov-based Manta ray foraging optimizer for global optimization and image segmentation. *Engineering Applications of Artificial Intelligence*, 98(May 2020): 104105. https://doi.org/10.1016/j.engappai.2020.104105.

Faramarzi, A., Heidarinejad, M., Mirjalili, S. and Gandomi, A.H. (2020). Marine Predators Algorithm: A nature-inspired metaheuristic. *Expert Systems with Applications*, 152: 113377. https://doi.org/10.1016/j.eswa.2020.113377.

Fathy, A., Rezk, H. and Yousri, D. (2020). A robust global MPPT to mitigate partial shading of triple-junction solar cell-based system using manta ray foraging optimization algorithm. *Solar Energy*, 207(June): 305–316. https://doi.org/10.1016/j.solener.2020.06.108.

Houssein, E.H., Emam, M.M. and Ali, A.A. (2021). Improved manta ray foraging optimization for multilevel thresholding using COVID-19 CT images. *Neural Computing and Applications*, 33(24): 16899–16919. https://doi.org/10.1007/s00521-021-06273-3.

Houssein, E.H., Ibrahim, I.E., Neggaz, N., Hassaballah, M. and Wazery, Y.M. (2021). An efficient ECG arrhythmia classification method based on Manta ray foraging optimization. *Expert Systems with Applications*, 181(April): 115131. https://doi.org/10.1016/j.eswa.2021.115131.

Houssein, E.H., Zaki, G.N., Diab, A.A.Z. and Younis, E.M.G. (2021). An efficient Manta Ray Foraging Optimization algorithm for parameter extraction of three-diode photovoltaic model. *Computers & Electrical Engineering*, 94(April): 107304. https://doi.org/10.1016/j.compeleceng.2021.107304.

Kapur, J.N., Sahoo, P.K. and Wong, A.K.C. (1985). A new method for gray-level picture thresholding using the entropy of the histogram. pp. 273–285. *In*: *Computer Vision Graphics Image Processing*. Elsevier.

Lang, C. and Jia, H. (2019). Kapur's entropy for color image segmentation based on a hybrid Whale Optimization Algorithm. *Entropy*, 21(3). https://doi.org/10.3390/e21030318.

Mirjalili, S. and Lewis, A. (2016). The whale optimization algorithm. *Advances in Engineering Software*, 95: 51–67. https://doi.org/10.1016/j.advengsoft.2016.01.008.

Mohammad-Azari, S., Bozorg-Haddad, O. and Chu, X. (2018). Shark smell optimization (SSO) algorithm. *Studies in Computational Intelligence*, 720: 93–103. https://doi.org/10.1007/978-981-10-5221-7_10.

Oliva, D., Cuevas, E., Pajares, G., Zaldivar, D. and Osuna, V. (2014). A multilevel thresholding algorithm using electromagnetism optimization. *Neurocomputing*, 139: 357–381.

Oliva, D., Cuevas, E., Pajares, G., Zaldivar, D. and Perez-Cisneros, M. (2013). Multilevel thresholding segmentation based on Harmony Search Optimization. *Journal of Applied Mathematics*. https://doi.org/10.1155/2013/575414.

Sankur, B., Sankur, B. and Sayood, K. (2002). Statistical evaluation of image quality measures. *Journal of Electronic Imaging*, 11(2): 206. https://doi.org/10.1117/1.1455011.

Tang, A., Zhou, H., Han, T. and Xie, L. (2021). A modified Manta ray foraging optimization for global optimization problems. *IEEE Access*, 9: 128702–128721. https://doi.org/10.1109/ACCESS.2021.3113323.

Upadhyay, P. and Chhabra, J.K. (2020). Kapur's entropy-based optimal multilevel image segmentation using Crow Search Algorithm. *Applied Soft Computing*, 97: 105522. https://doi.org/10.1016/j.asoc.2019.105522.

Zaldívar, D., Morales, B., Rodríguez, A., Valdivia-G, A., Cuevas, E. and Pérez-Cisneros, M. (2018). A novel bio-inspired optimization model based on Yellow Saddle Goatfish behavior. *BioSystems*, 174(September): 1–21. https://doi.org/10.1016/j.biosystems.2018.09.007.

Zhao, W., Zhang, Z. and Wang, L. (2020). Manta ray foraging optimization: An effective bio-inspired optimizer for engineering applications. *Engineering Applications of Artificial Intelligence*, 87(October 2019): 103300. https://doi.org/10.1016/j.engappai.2019.103300.

CHAPTER 5

Archimedes Optimization Algorithm and Cross-entropy

5.1 Archimedes Optimization Algorithm

In this chapter, we will learn about the Archimedes Optimization Algorithm and the cross-entropy. Among the physics-based metaheuristics, we can find many approaches taking its metaphor on physical phenomena or laws. A recent example of such a category is the Archimedes optimization algorithm (AOA). It was initially proposed by Hashim et al. 2021. As a source of inspiration, the algorithm uses the displacement of fluid while an object is immersed, and this phenomenon produces an upward force that resists the object's weight, also known as buoyant force. See Figure 5.1. From a computational perspective, in AOA, the candidate solutions are considered objects immersed in a fluid.

In the field of energy, the AOA has been applied to a wide range of problems. Zhang et al. (2021) used the AOA for accurately predicting wind speed for Eolic energy generation by proposing an ensemble forecasting system that takes a multi-objective version of the AOA (Zhang et al., 2021). Fathy et al. (2021) simulated the maximum power point tracker of a wind energy generation system where the AOA helped them to tune the converter duty cycle to maximize the energy generation (Fathy et al., 2021). She

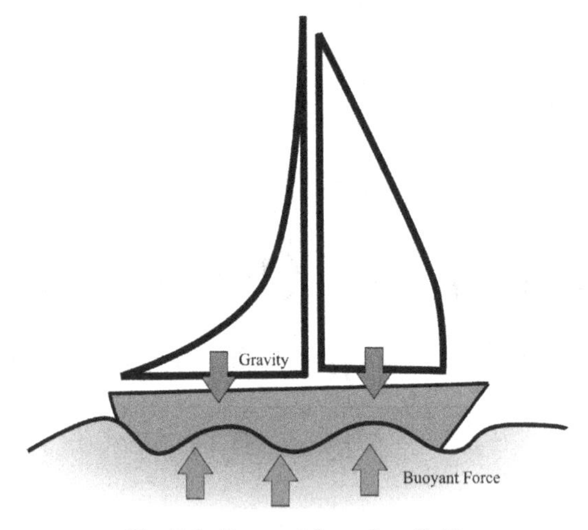

Fig. 5.1. Buoyant force in a fluid.

et al. (2021) used a variant of the AOA to optimize the energy consumption of buildings, as this problem has been gaining relevance during the last years (She et al., 2021). Yao and Hayati (2021) applied the AOA to find optimal design parameters of a Proton Exchange Membrane fuel cell with excellent results (Yao and Hayati, 2021). Even more, the problem of finding the optimal allocation of renewable energy sources inside a distribution network is addressed by Eid and El-Kishky (2021).

Another field where the AOA has been applied is machine learning. Desuky et al. (2021) proposed an enhanced AOA for feature selection in real-world datasets (Desuky et al., 2021). The AOA has also been applied in the fields of computer vision and deep learning. Anrrose et al. (2021) presented a Cloud-based Platform for Soybean Plant Disease Classification Using Archimedes Optimization Based Hybrid Deep Learning Model, which uses a mixture of wavelet packet decomposition and the architecture long short-term memory (Annrose et al., 2021).

5.1.1 Initialization

As most population-based stochastic optimization algorithms, the first population of objects O is randomly initialized within the upper u and lower l bounds of the problem according to the formula given in Eq. 5.1:

$$O_i = l_i + rand \times (u_i - l_i), i = 1,2,...,N \tag{5.1}$$

where the subindex i indicates which of the N objects O is considered. The other two elements of the algorithm are the D-dimensional vectors density den_i and volume vol_i, which are initialized randomly with values ranging from 0 to 1.

As the objects interact in the fluid, the acceleration of each object is relevant in this approach. During its initialization, it is calculated as a random value between the bounds of the problem according to Eq. 5.2.

$$acc_i = lb_i + rand \times (ub_i - lb_i) \tag{5.2}$$

5.1.2 Update

After the initialization of objects immersed in a fluid, the fitness function is evaluated for the entire population, and the best object is retained as x_{best}; also, its density den_{best}, volume vol_{best}, and acceleration acc_{best} are maintained to help the updating process.

The update equations of the density and volume are designed to move such values towards the best object found at that iteration t. This mechanism will produce many collisions as the algorithms try to reach an equilibrium (Eqs. 5.3 and 5.4).

$$den_i^{t+1} = den_i^t + rand \times \left(den^{best} - den_i^t\right) \tag{5.3}$$

$$vol_i^{t+1} = vol_i^t + rand \times \left(vol^{best} - vol_i^t\right) \tag{5.4}$$

The collisions are handled by a transfer operator designed to provide a smooth transition between exploration and exploitation phases by considering a maximum number of iterations $t_{\max}$ (Eq. 5.5)

$$TF = \exp\left(\frac{t - t_{\max}}{t_{\max}}\right) \quad (5.5)$$

In this scheme, the *TF* value slowly increases until it reaches 1. To balance the search process, also a decreasing density factor is included in the approach and is formulated as given in Eq. 5.6:

$$d^{t+1} = \exp\left(\frac{t - t_{\max}}{t_{\max}}\right) - \left(\frac{t}{t_{\max}}\right) \quad (5.6)$$

5.1.3 Exploration phase

During the exploration phase, the acceleration is calculated considering as a reference a random object denoted as random material *mr*. The exploration occurs when the transfer operator is 0.5 (Eq. 5.7).

$$acc_i^{t+1} = \frac{den_{mr} + vol_{mr} \times acc_{mr}}{den_i^{t+1} \times vol_i^{t+1}} \quad (5.7)$$

In both exploration and exploitation, it is crucial to normalize the acceleration to have a significant value that will translate into a mechanism that will generate large accelerations when the object is far from its reference, according to the formula shown in Eq. 5.8.

$$acc_{i-norm}^{t+1} = u \times \left(\frac{acc_i^{t+1} - \min(acc)}{\max(acc) - \min(acc)}\right) + l \quad (5.8)$$

where l and u are constants that normalize the range. Here 0.1 and 0.9 are assigned, respectively.

Then, the position is updated considering the previous concepts and a constant C_I given by the user (Eq. 5.9).

$$x_i^{t+1} = x_i^t + C_1 \times rand \times acc_{i-norm}^{t+1} \times d \times \left(x_{rand} - x_i^t\right) \quad (5.9)$$

5.1.4 Exploitation phase

In this phase, the collision between objects is not considered. The program will be in exploitation when the transfer operator is greater than 0.5 and will produce an acceleration given by the Eq. 5.10 where the best solution is taken as reference.

$$acc_i^{t+1} = \frac{den_{best} + vol_{best} \times acc_{best}}{den_i^{t+1} \times vol_i^{t+1}} \quad (5.10)$$

The updating process during the exploitation phase also takes a constant value given by the user C_2, but also takes a parameter T (Eq. 5.11) associated with the transfer factor TF ($T = C3 \times TF$)

$$x_i^{t+1} = x_i^t + F \times C_2 \times rand \times acc_{i-norm}^{t+1} \times d \times \left(T \times x_{best} - x_i^t\right) \quad (5.11)$$

where F is a flag incorporated to determine the direction of movement of the object and is calculated as given in the following Eq. 5.12:

$$F = \begin{cases} +1 P \leqslant 0.5 \\ -1 P > 0.5 \end{cases} \quad (5.12)$$

where P is randomly selected by the user.

Figure 5.2 presents a visual representation of the algorithm. It is possible to follow the equations to replicate the results of the authors.

5.2 Cross-Entropy

Kullback proposed the cross-entropy in 1968.. Let $\mathbf{J} = \{j_1, j_2, \ldots jN)$ and $\mathbf{G} = \{g_1, g_2, \ldots gN)$ be two probability distributions on the same set. The cross-entropy between F and G is an information-theoretic distance between the two distributions, and it is defined as shown in Eq. 5.13:

$$D(\mathbf{J}, \mathbf{G}) = \sum_{i=1}^{N} j_i \log \frac{j_i}{g_i} \quad (5.13)$$

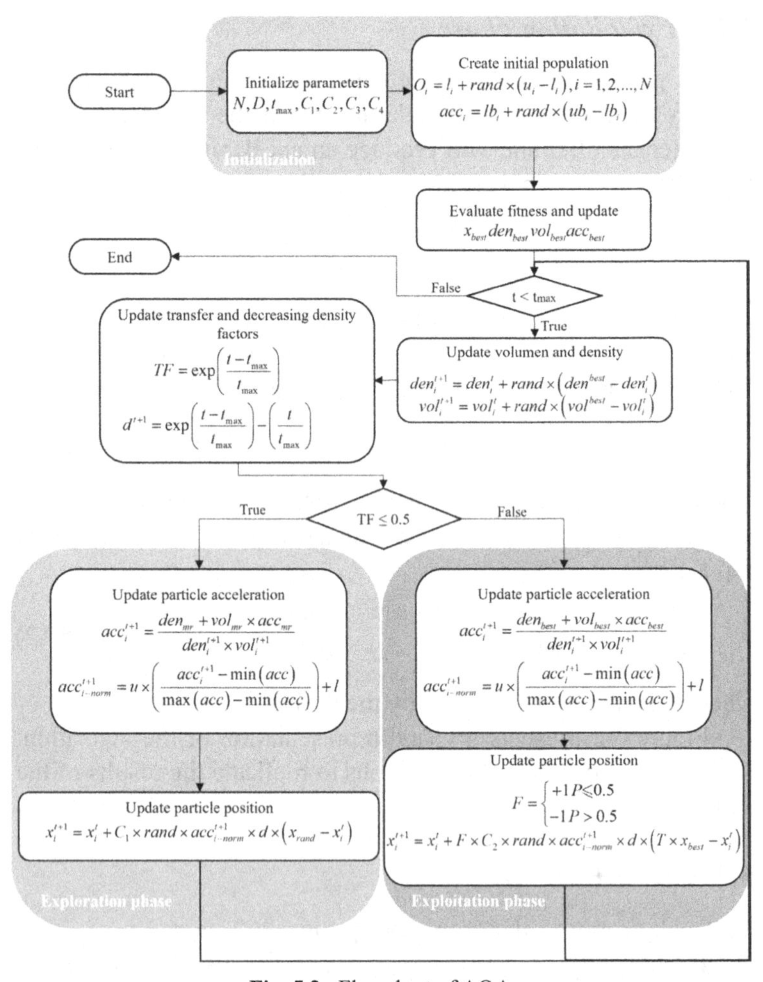

Fig. 5.2. Flowchart of AOA.

The minimum cross-entropy thresholding (MCET) algorithm (Li and Lee, 1993) selects the threshold by minimizing the cross-entropy between the threshold version and its original image. The original image is **I,** and $h^{Gr}(i)$, $I = 1, 2, \ldots, L$ is the corresponding histogram with L being the number of gray levels. Then the threshold image, denoted by $\mathbf{I}_t$ using th as the threshold value, is constructed by Eq. 5.14:

$$\mathbf{I}_t(x,y) = \begin{cases} \mu(1,th), & \mathbf{I}(x,y) < th, \\ \mu(th,L+1), & \mathbf{I}(x,y) \geq th, \end{cases} \tag{5.14}$$

where:

$$\mu(a,b) = \sum_{i=a}^{b-1} ih^{Gr}(i) / \sum_{i=a}^{b-1} h^{Gr}(i) \tag{5.15}$$

Since Eq. 5.15 generates a threshold image instead of an entropy value, the cross-entropy is rewritten as an objective function (Eq. 5.16):

$$f_{Cross}(th) = \sum_{i=1}^{th-1} ih^{Gr}(i)\log\left(\frac{i}{\mu(1,th)}\right) + \sum_{i=th}^{L} ih^{Gr}(i)\log\left(\frac{i}{\mu(th,L+1)}\right) \tag{5.16}$$

The objective function considers a single threshold value for bilevel thresholding. Eq. 5.16 can be extended to a multilevel approach. First Eq. 5.16 can be expressed as (Eq. 5.17):

$$f_{Cross}(th) = \sum_{i=1}^{L} ih^{Gr}(i)\log(i) - \sum_{i=1}^{th-1} ih^{Gr}(i)\log(\mu(1,th)) - \sum_{i=th}^{L} ih^{Gr}(i)\log(\mu(th,L+1)) \tag{5.17}$$

The multilevel approach is based on the use of the vector $\mathbf{th} = [th_1, th_2, \ldots, th_{nt}]$, which contains nt different thresholds values (Eq. 5.18).

$$f_{Cross}(\mathbf{th}) = \sum_{i=1}^{L} ih^{Gr}(i)\log(i) - \sum_{i=1}^{nt} H_i \tag{5.18}$$

where k is the number of threshold and entropies to calculate as given in Eq. 5.19.

$$H_1 = \sum_{i=1}^{th_1-1} ih^{Gr}(i)\log(\mu(1,th_1)) \tag{5.19}$$

$$H_k = \sum_{i=th_{k-1}}^{th_k-1} ih^{Gr}(i)\log(\mu(th_{k-1},th_k)), \quad 1 < k < nt$$

$$H_{nt} = \sum_{i=th_{nt}}^{L} ih^{Gr}(i)\log(\mu(th_{nt},L+1))$$

It must be noticed that the MCET process can easily be implemented for a color image. For this purpose, each channel of the picture is treated as a single gray level image, allowing to compute the cross-entropy as explained in this section.

5.3 Multilevel Thresholding with AOA and Cross-entropy

This section introduces the implementation of the AOA for multilevel image thresholding by using the cross-entropy formulation. The solutions contain decision variables that correspond to a set of thresholds. The population of objects $\boldsymbol{O}$ in the AOA is defined as:

$$\mathbf{O} = [O_1, O_2, O_3, \ldots, O_N],\ O_i = [th_1, th_2, \ldots, th_k]^T \quad (5.20)$$

Figure 5.3 shows the flowchart of the AOA implementation for digital image thresholding by using the cross-entropy formulation. The stop criterion in Figure 5.3 is the maximum number of iterations (*Maxiter*). When the number of iterations

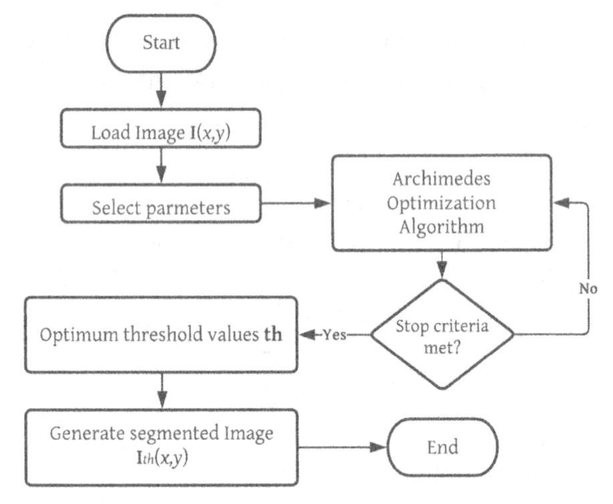

Fig. 5.3. Flowchart of AOA cross-entropy.

reaches the maximum value, the algorithm ends, and the best set of thresholds is extracted and applied to the orifical image in grayscale. As discussed in previous chapters, this procedure could be extended for color images as RGB by using the method over each channel separately.

5.4 Experiments

The implementation of the AOA for image thresholding previously described has been tested over a set of benchmark images as described in Chapter 2. The AOA was configured in the standard version suggested by the authors in (Hashim et al., 2021). For the thresholding process, four different values $th = 2, 4, 8, 16$ were applied. Such amounts of thresholds help to verify the performance of the AOA in multidimensional problems. The AOA belongs to the class of stochastic optimization algorithms. Due to its stochastic nature, it is not possible to judge the performance of the algorithm over a single execution, so the experiments contain 35 independent runs over each image using every threshold value. Besides, the segmented images are also analyzed by using the quality metrics, Peak Signal to Noise Ratio (PSNR) (Sankur et al., 2002), the Structural Similarity Index (SSIM), and the Feature Similarity Index (FSIM) that were described in Table 1.1 in Chapter 1. The segmented images are presented in Figure 5.4.

We can observe interesting elements. For example, in the image gral_04, the sky and the bridge share the same class despite having clearly differentiated structures. In most cases, the level of detail in the image largely grows as the number of thresholds increases; however, from 8 thresholds to 16, the difference is only noticeable over a close-up. This behavior can be exemplified by the portrait on the image gral_02.

Regarding the quality of the outputs, Table 5.1 presents the results generated by the output images in terms of the PSNR, SSIM, and FSIM by using the AOA in combination with the cross-entropy fitness function.

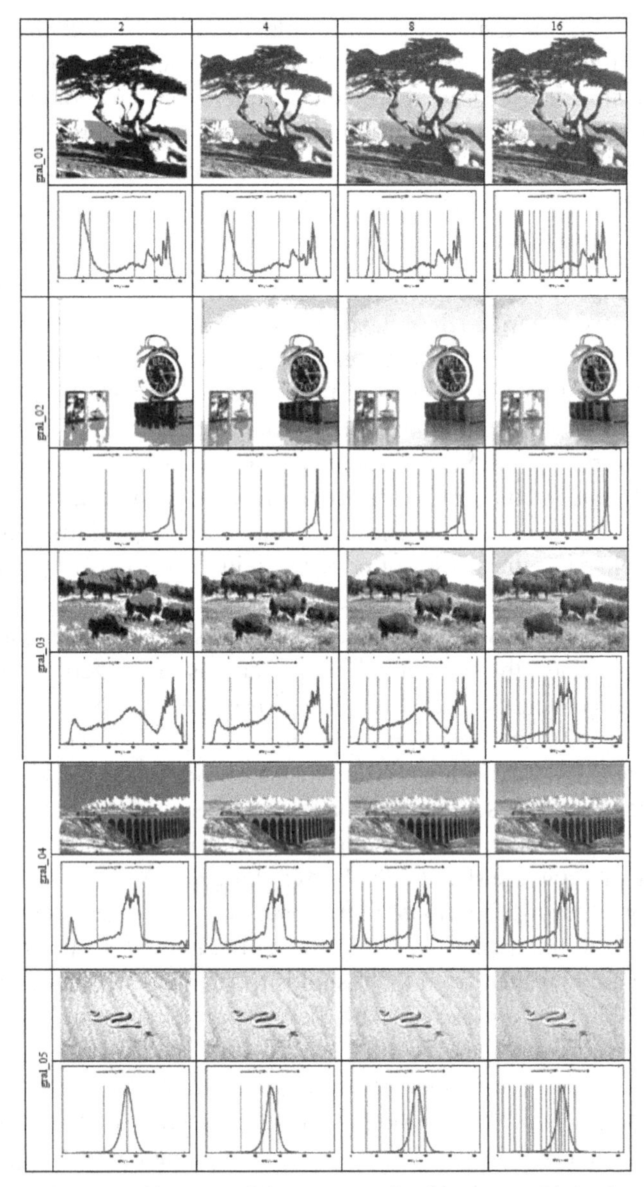

Fig. 5.4. Segmented images and the corresponding histogram with the thresholds obtained by using the AOA with the cross-entropy.

Table 5.1. PSNR, SSIM, and FSIM values on general benchmark images using cross-entropy.

Image	nTh	AOA cross-entropy					
		PSNR		SSIM		FSIM	
		mean	std	mean	std	mean	std
gral_01.tiff	2	14.2264	0.00E+00	0.5902	0.00E+00	0.7731	0.00E+00
	4	17.2817	7.40E-02	0.7033	4.60E-03	0.8362	2.10E-03
	8	22.8794	5.01E-01	0.8991	1.69E-02	0.9006	1.12E-02
	16	29.5237	6.52E-01	0.9695	9.90E-03	0.9646	8.90E-03
gral_02.tiff	2	14.2155	0.00E+00	0.8224	0.00E+00	0.7960	0.00E+00
	4	19.7716	1.25E-01	0.8800	5.30E-03	0.8262	6.00E-03
	8	24.7333	2.95E-01	0.9348	5.10E-03	0.8853	3.30E-03
	16	30.9992	3.67E-01	0.9754	2.70E-03	0.9489	4.30E-03
gral_03.jpg	2	13.9968	0.00E+00	0.6331	0.00E+00	0.6575	0.00E+00
	4	18.6544	1.02E-01	0.8130	9.00E-04	0.8259	1.80E-03
	8	23.7654	1.13E-01	0.9080	3.80E-03	0.9125	1.10E-03
	16	29.4720	1.38E-01	0.9729	3.30E-03	0.9669	3.60E-03
gral_04.jpg	2	12.8436	0.00E+00	0.6969	0.00E+00	0.6860	0.00E+00
	4	19.5837	6.10E-02	0.8543	1.10E-03	0.8107	9.00E-04
	8	23.9651	1.67E-01	0.9288	4.00E-03	0.8968	4.60E-03
	16	29.5115	3.75E-01	0.9785	1.40E-03	0.9577	4.70E-03
gral_05.tiff	2	18.9865	0.00E+00	0.7120	0.00E+00	0.7484	0.00E+00
	4	25.4252	3.95E-01	0.9080	8.90E-03	0.8826	8.70E-03
	8	30.3172	5.90E-01	0.9672	4.90E-03	0.9486	5.20E-03
	16	34.5787	4.81E-01	0.9867	1.40E-03	0.9771	3.50E-03

Table 5.1 presents the mean and the standard deviation (std) of the PSNR, SSIM, and FSIM values obtained by comparing the original grayscale image with the segmented image obtained by the thresholds computed using the EO. As in previous chapters, the results in the table are computed from a set of 35 independent experiments. The std permits to see the stability of the algorithm; in lower dimension as in *th* = 2 the std values are closer to 0 and it means that the PO is more stable with the results. Meanwhile, for *th* = 16, the EO presents some fluctuations that are reflected in a higher std. Regarding the mean values of the quality metrics, for all of them, when the number of thresholds increases, the values also increase. For PSNR, SSIM, and FSIM, it represents a higher similarity between the two images. Specifically, for SSIM and FSIM, when its values are closer to 1, the internal structure and features of the pixels from the segmented image are closer to the original image.

We can observe that the gral_04 image seems to be the most challenging as the PSNR recorded for two thresholds is much lower in comparison to the other images with the same number of thresholds. The image gral_05 surprisingly shows high values over the entire set of metrics, and this can be produced due to the simplicity of the image histogram.

5.5 Conclusions

The cross-entropy has been applied to several research articles over the last few decades. It provides a robust entropic criterion based on the divergence of two probability distributions. The popularity of this approach is due to its simplicity; also, the code is available in different repositories on the web.

The AOA was proposed recently, and is gaining acceptance in many areas. In the case of image segmentation, this method is a suitable alternative since it is easy to adapt any nonparametric criterion, in this case, cross-entropy.

Exercises

5.1 Implement the cross-entropy formulation for a single threshold presented in Eq. 5.18 in any programming language.

5.2 Use the exhaustive search to find the best threshold of the cameraman image using the cross-entropy for a single threshold.

5.3 Program the cross-entropy for multilevel thresholding presented in Eq. 5.18.

5.4 Implement the AOA in combination with the cross-entropy for multilevel thresholding and test it over different images.

5.5 Program the PSNR, SSIM, and FSIM and verify the quality of the segmentation.

5.6 Implement the program created in Exercise 5.4 for color images.

References

Annrose, J., Rufus, N.H.A., Rex, C.R.E.S. and Immanuel, D.G. (2021). A cloud-based platform for soybean plant disease classification using archimedes optimization-based hybrid deep learning model. *Wireless Personal Communications* 2021: 1–23. https://doi.org/10.1007/S11277-021-09038-2.

Desuky, A.S., Hussain, S., Kausar, S., Islam, M.A. and Bakrawy, L.M.E. (2021). EAOA: An enhanced archimedes optimization algorithm for feature selection in classification. *IEEE Access*, 9: 120795–120814. https://doi.org/10.1109/ACCESS.2021.3108533.

Eid, A. and El-Kishky, H. (2021). Multi-objective archimedes optimization algorithm for optimal allocation of renewable energy sources in distribution networks. *Lecture Notes in Networks and Systems*, 211 LNNS: 65–75. https://doi.org/10.1007/978-3-030-73882-2_7.

Fathy, A., Alharbi, A.G., Alshammari, S. and Hasanien, H.M. (2021). Archimedes Optimization Algorithm based maximum power point tracker for wind energy generation system. *Ain Shams Engineering Journal*. https://doi.org/10.1016/J.ASEJ.2021.06.032.

Hashim, F.A., Hussain, K., Houssein, E.H., Mabrouk, M.S. and Al-Atabany, W. (2021). Archimedes optimization algorithm: A new metaheuristic algorithm for solving optimization problems. *Applied Intelligence*, 51(3): 1531–1551. https://doi.org/10.1007/S10489-020-01893-Z/TABLES/14.

Kullback, S. (1968). *Information Theory and Statistics* (Dover Books on Mathematics).

Li, C.H. and Lee, C.K. (1993). Minimum cross-entropy thresholding. *Pattern Recognition*, 26(4): 617–625. https://doi.org/10.1016/0031-3203(93)90115-D.

Sankur, B., Sankur, B. and Sayood, K. (2002). Statistical evaluation of image quality measures. *Journal of Electronic Imaging*, 11(2): 206. https://doi.org/10.1117/1.1455011.

She, C., Jia, R., Hu, B.-N., Zheng, Z.-K., Xu, Y.-P. and Rodriguez, D. (2021). Life-cycle cost and life-cycle energy in zero-energy building by multi-objective optimization. *Energy Reports*, 7: 5612–5626. https://doi.org/10.1016/J.EGYR.2021.08.198.

Yao, B. and Hayati, H. (2021). Model parameters estimation of a proton exchange membrane fuel cell using improved version of Archimedes Optimization Algorithm. *Energy Reports*, 7: 5700–5709. https://doi.org/10.1016/J.EGYR.2021.08.177.

Zhang, L., Wang, J., Niu, X. and Liu, Z. (2021). Ensemble wind speed forecasting with multi-objective Archimedes Optimization Algorithm and sub-model selection. *Applied Energy*, 301: 117449. https://doi.org/10.1016/J.APENERGY.2021.117449.

Chapter 6

Equilibrium Optimizer and Masi Entropy

6.1 Equilibrium Optimizer

In this chapter, we will learn about the Equilibrium Optimizer and an entropy formulation known as Masi and how they can interact to solve the image thresholding problem. The Equilibrium Optimizer (EO) is a stochastic optimization algorithm inspired by physical laws. The main idea of the EO is finding a dynamic mass balance on a limited volume. This is achieved by a mass balance equation designed to evaluate solutions that find an equilibrium state (Faramarzi et al., 2020). The EO has been widely used over areas such as energy generation and distribution. For instance, in the article by Abdel-Basset et al. (2020), they use an improved version of the EO to estimate parameters of solar panels (Abdel-Basset et al., 2020). Since the parameter identification of models is quite popular in the field of energy, we can find examples of the EO applied not only to solar cells but also to fuel cells; the work by Seleem et al. (2021) is a good example (Seleem et al., 2021). On a similar note, Abdul-Hamied et al. (2020) used the EO to address the optimal power flow problem over hybrid AC/DC distribution grids (Abdul-Hamied et al., 2020).

Another type of application where the EO has been applied is machine learning, specifically in the problem of feature selection. Gao et al. (2020) used binary versions of the EO for selecting an optimal set of features for a given classification problem (Gao et al., 2020). Too and Mirjalili (2020) developed a General Learning modification for the EO; their approach is applied to feature selection over biological data (Too and Mirjalili, 2020).

In the related literature, we can find examples of implementations of EO versions to the problem of image processing with different perspectives and objectives. Wunnava et al. (2020) presented a modified version of the EO which takes as the objective function the concept of shred boundary instead of entropic measures over the histogram (Wunnava et al., 2020). Then, an opposition-based Laplacian variation of the EO was proposed to address general image thresholding using Otsu's method (Dinkar et al., 2021). Naik et al. (2021) presented an exciting formulation of a context-sensitive entropy dependency for remote sensing image thresholding (Naik et al., 2021).

We can find other types of problems such as image registration, where the main idea is transforming different images of the same scene into a single coordinate system; the approach presented by Gui et al. (2021) exemplifies this issue over medical images using a modification of the EO that dynamically updates the control parameters (Gui et al., 2021). Following medical applications, we can find the contribution of Dinh, where a multimodal medical image fusion is described (Dinh, 2021).

The EO has been applied to a large number of applications of classical image processing problems. Also, there are implementations of the EO designed to assist a wide range of applications, such as Deep Learning approaches, aerial vehicle path planning. Lan et al. (2021) proposed an EO version that optimizes the parameters of a long-short term memory (LSTM) network (Lan et al., 2021). Tang et al. (2021) used the EO for planning aerial routes for unmanned aerial vehicles (Tang et al., 2021).

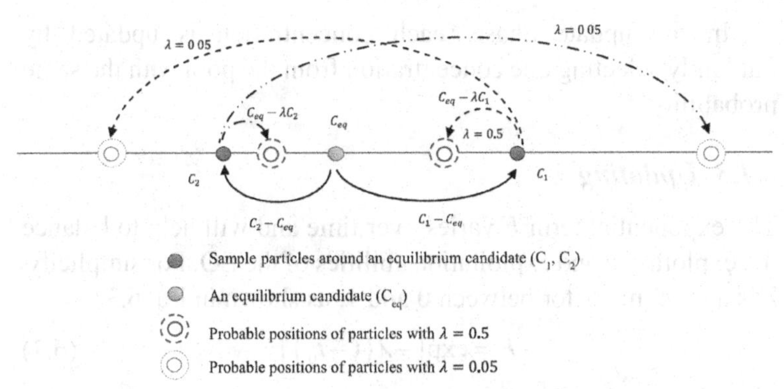

Fig. 6.1. 1-D presentation of concentrations updating aid in exploration and exploitation (Faramarzi et al., 2020).

According to its original formulation (Faramarzi et al., 2020), the general model of the EO algorithm can be summarized in the following points, as shown in Figure 6.1.

6.1.1 Initializing

The initial concentration is based on the number of particles and dimensions with random initialization in the search space as depicted in Eq. 6.1.

$$C_i^{initial} = C_{min} + rand_i \left(C_{max} - C_{min} \right) \qquad i = 1,2,\ldots..n \quad (6.1)$$

6.1.2 Pooling

EO uses an exciting mechanism called pooling to manage the social contribution of each concentration (solution). The four best candidates are selected and contained in a vector and the average of the four of them is taken. The final structure comprises five candidate concentrations where the first four will promote the exploration while the average concentration will contribute to the exploitation (Eq. 6.2).

$$\vec{C}_{eq.pool} = \{\vec{C}_{eq(1)}, \vec{C}_{eq(2)}, \vec{C}_{eq(3)} \vec{C}_{eq(4)}, \vec{C}_{eq(avg)}\} \qquad (6.2)$$

In the update phase, each concentration is updated by randomly selecting one concentration from the pool with the same probability.

6.1.3 Updating

The exponential term F varies over time and will help to balance the exploration and exploitation abilities of the EO. For simplicity, λ is a random vector between 0 and 1, as shown in Eq. 6.3.

$$F = \exp\left[-\lambda\left(\mathrm{t} - t_0\right)\right] \tag{6.3}$$

where t is a function that decreases over time according to the current iteration $Iter$ and the total number of iterations Max_iter and is controlled by a constant a_2 that limits the exploitation ability of the algorithm as given in Eq. 6.4.

$$t = \left(1 - \frac{Iter}{Max\_iter}\right)^{\left(a_2 \frac{Iter}{Max\_iter}\right)} \tag{6.4}$$

Equation 6.5 shows the other component of the exponential term F:

$$\vec{t}_0 = \frac{1}{\vec{\lambda}} \ln\left(-a_1 sign(\vec{r} - 0.5)\left[1 - e^{-\vec{\lambda}t}\right]\right) + t \tag{6.5}$$

Similar to t, a_1, is a tunning constant that governs the exploration of the algorithm, $\vec{r}$ is randomly generated vector within 0 and 1 while the function sign will indicate the direction of the search for a given dimension.

One key element of the EO is the use of a generation rate G that improves the exploitation phased given by Eq. 6.6:

$$\vec{G} = \vec{G}_0 e^{-\vec{\lambda}(t - t_0)} = \vec{G}_0 \vec{F} \tag{6.6}$$

Here $\vec{G}_0$ indicates the initial value as defined by Eq. 6.7:

$$\overrightarrow{G_0} = \overrightarrow{GCP}\left(\overrightarrow{C_{eq}} - \vec{\lambda}\vec{C}\right) \tag{6.7}$$

$$\overrightarrow{GCP} = \begin{cases} 0.5r_1 & r_2 \geq GP \\ 0 & r_2 < GP \end{cases} \tag{6.8}$$

where the Generation rate Control Parameter GCP indicates the possibility of the term's contribution to the updating process while r_1 and r_2 are random numbers drawn from a uniform distribution between 0 and 1 (Eq. 6.8); this is achieved by considering the generation probability *GP*. In the original paper, the authors suggested that 0.5 in *GP* provides a good balance in the dynamics of the algorithm.

Finally, the EO will update each concentration as follows (Eq. 6.9):

$$C = C_{eq} + \left(C_0 - C_{eq}\right)F + \frac{G}{\lambda V}(1 - \mathrm{F}) \tag{6.9}$$

where for simplicity V = 1.

6.2 Masi Entropy

A way to measure the amount of information that a random variable stores is known as entropy; in the context of information theory, proposed by the mathematician Claude E. Shannon (Mahmoudi and El Zaart, 2012), a good segmentation should maximize the uniformity of pixels within each segmented region and minimize the uniformity across the region. Entropy has been used as a metric to perform segmentation with outstanding results. The method proposed by Kapur is used for determining the ideal thresholding image segmentation; this algorithm was developed for bi-level thresholding (Kapur et al., 1985). The method can be extended to solve multilevel thresholding problems as follows:

A digital image in grayscale I can be defined by a matrix size $M \times N$ where $I = \{l_{ij}, 1 \leq i \leq M, 1 \leq j \leq N\}$ then l_{ij} is the gray level of the image in the pixel (i, j). These gray levels L are in the range $\{1, 2..., L - 1\}$. Then one can define $P_i = h(i)/D$ $(1 \leq i \leq L - 1)$

```
1.  Load image I(x, y)
2.  Calculate the gray-scale histogram h^gr(I)
3.  Initialize the population TH of particles th_i(i=1,2,3,...,n) (Eq. 6.1)
4.  Set fitness value of the four particles in equilibrium pool, th_eq, a large value
5.  Set parameter's value a_1 = 1; a_2 = 2; GP = 0.5;
6.  while ( iter < maxIter)
7.          for each i particle
8.                  Calculate fitness values for each particle (Eq. 6.10)
9.                  if (f(th_i) < f(th_eq(1)))
10.                         Set th_eq(1) with th_i and f(th_eq(1)) with f(th_i)
11.                 elseif (f(th_i) > f(th_eq(1)) and f(th_i) < f(th_eq(2)))
12.                         Set th_eq(2) with th_i and f(th_eq(2)) with f(th_i)
13.                 elseif (f(th_i) > f(th_eq(2)) and f(th_i) < f(th_eq(3)))
14.                         Set th_eq(3) with th_i and f(th_eq(3)) with f(th_i)
15.                 elseif (f(th_i) > f(th_eq(3)) and f(th_i) < f(th_eq(4)))
16.                         Set th_eq(4) with th_i and f(th_eq(4)) with f(thc_i)
17.                 end if
18.         end for
19.         th_eq(avg) = (th_eq(1) + th_eq(2) + th_eq(3) + th_eq(4))/4
20.         The equilibrium pool th_eq,pool = [th_eq(1), th_eq(2), th_eq(3), th_eq(4), th_eq(avg)]
21.         Accomplish the memory saving
22.         Assign t (Eq. 6.4)
23.         for each i particle
24.                 Choose one candidate from C_eq,pool randomly
25.                 Generate two vectors, namely r, λ randomly
26.                 Construct F (Eq. 6.3)
27.                 Construct GCP (Eq. 6.8)
28.                 Construct G_0 (Eq. 6.7)
29.                 Construct G (Eq. 6.6)
30.                 Update the concentration (Eq. 6.9)
31.         end for
32.         update the equilibrium pool if there is particle better
33.         iter = iter + 1
34. end while
35. Select the best solution th_best and construct the segmented image I_th(x, y) (Eq. 1.2)
```

Fig. 6.2. Pseudocode of EO for image thresholding.

where; $h(i)$ is the number of pixels with gray level i. Moreover, D denotes the total number of pixels in the image.

Having a problem to select nt thresholds $[th_1, th_2, \ldots, th_{nt}]$, for an image I, the objective function to maximize is shown in Eqs. 6.10 and 6.11:

$$\mathbf{th}^* = \arg\max\left(\sum_{i=0}^{nt} H_i\right) \tag{6.10}$$

where

$$\begin{aligned}
\omega_0 &= \sum_{i=0}^{th_1-1} P_i, \quad H_0 = -\sum_{i=0}^{th_1-1} \frac{P_i}{\omega_0} \ln \frac{P_i}{\omega_0} \\
\omega_1 &= \sum_{i=th_1}^{th_2-2} P_i, \quad H_1 = -\sum_{i=th_1}^{th_2-1} \frac{P_i}{\omega_1} \ln \frac{P_i}{\omega_1} \\
&\vdots \\
\omega_n &= \sum_{i=th_{nt}}^{th_{nt}-1} P_i, \quad H_{nt} = -\sum_{i=th_{nt}}^{th_{nt}-1} \frac{P_i}{\omega_{nt}} \ln \frac{P_i}{\omega_{nt}}
\end{aligned} \tag{6.11}$$

6.2.1 *Masi entropy*

Despite the entropy formulation proposed by Kapur that can effectively be used for image segmentation, other entropy-based approaches have been prosed, including the formulations of Rényi (Li et al., 2007) and Tsallis (Tsallis, 1988). However, the properties of the distribution of the histogram can benefit one entropy formulation over another. To overcome this problem, the entropy Masi was proposed in the article, *A step beyond Rényi and Tsallis entropies* as a generalization of the Rényi and Tsallis entropies and proposed by Marco Masi (Masi, 2005). This generalization takes Rényi's additive and Tsali's non-extensive propriety. This entropy for a complete probability distribution $P = \{p_1,\ldots,p_i,\ldots,p_n\}$, $0 \le p_i \le 1$, $i = 1,\ldots, n$, $\sum_{i=1}^{n} p_1$, is defined by the Eq. 6.11:

$$S_r = \frac{1}{1-r} \log\left[1-(1-r)\sum_{i=1}^{nt} p_i \log p_i\right] \tag{6.12}$$

where $r > 0$ and $r \neq 1$. The value of r is the measure of additivity/non-extensivity propriety of Tsalli's and Rényi's entropies. The value of r is commonly configured manually, but in this article, it is also optimized for each image.

Equation 6.12 shows the Masi formulation for multilevel thresholding with *n* thresholds:

$$E_{rT} = E_{rT_0} + E_{rT_1} + \ldots + E_{rT_n} \tag{6.13}$$

where

$$\begin{aligned}
E_{rT0} &= \frac{1}{1-r}\log\left[1-(1-r)\sum_{i=0}^{th_1-1}\left(\frac{h_i}{\omega_0}\right)\log\left(\frac{h_i}{\omega_0}\right)\right], \omega_0 = \sum_{i=0}^{th_1-1} h_i \\
E_{rT1} &= \frac{1}{1-r}\log\left[1-(1-r)\sum_{i=th_1}^{th_2-1}\left(\frac{h_i}{\omega_1}\right)\log\left(\frac{h_i}{\omega_1}\right)\right], \omega_0 = \sum_{i=th_1}^{th_2-1} h_i \\
E_{rTn} &= \frac{1}{1-r}\log\left[1-(1-r)\sum_{i=th_n+1}^{L-1}\left(\frac{h_i}{\omega_n}\right)\log\left(\frac{h_i}{\omega_n}\right)\right], \omega_n = \sum_{i=th_n}^{L-1} h_i
\end{aligned} \tag{6.14}$$

In Eq. 6.13, E_{rTi} indicates the entropy value of the *i*-th class of the thresholding problem. The general idea is to identify the best set of thresholds **th** that maximizes the value or the entropy according to Eq. (6.14). Formally, the multilevel thresholding problem can be stated as:

$$\mathbf{th}^* = \arg\max\left(\sum_{i=0}^{nt} E_{rTi}\right) \tag{6.15}$$

In his generalization, Masi proposes a method to obtain the no extensivity from Rényi's Entropy, and the additive propriety of Tsallis entropy only by modifying a value *r* in the proposed formula. Since each image has a different distribution, an entropy formulation can be better than others for that specific image; with the use of Masi entropy, we improve the influence of two entropies Rényi and Tsallis, to obtain a better segmentation.

6.3 Multilevel Thresholding with EO and Masi Entropy

This section introduces the implementation of the EO for multilevel image thresholding by using the Masi entropy formulation. The solutions contain decision variables that correspond to a set of thresholds. The population of concentrations (solutions) $\boldsymbol{C}$ in the EO is defined in Eq. 6.15 as:

$$\mathbf{C} = [C_1, C_2, C_3, \ldots, C_N], \ C_i = [th_1, th_2, \ldots, th_k]^T \quad (6.16)$$

Figure 6.3 shows the flowchart of the EO implementation for digital image thresholding by using Masi's entropy formulation. The stop criterion in Figure 6.3 is the maximum number of iterations (*Maxiter*). When *It* reaches the maximum value, the algorithm ends, and the best set of thresholds is extracted and applied to the original image in grayscale Eq. 1.2. As discussed in previous chapters, this procedure could be extended for color images as RGB by using the method over each channel separately.

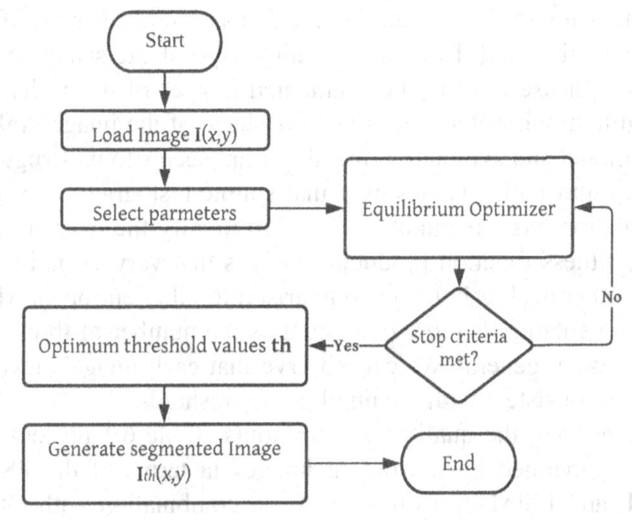

Fig. 6.3. Flowchart of EO-Masi-entropy.

6.4 Experiments

The implementation of the EO for image thresholding previously described has been tested over a set of benchmark images described in Chapter 2. The EO was configured in the standard version suggested by the authors (Faramarzi et al., 2020). For the thresholding process, four different values th = 2, 4, 8, 16 were applied. Such amounts of thresholds help to verify the performance of the EO in multidimensional problems. The EO belongs to the class of stochastic optimization algorithms. Due to its stochastic nature, it is not possible to judge the performance of the algorithm over a single execution, so the experiments contain 35 independent runs over each image using every threshold value. Besides, the segmented images are also analyzed by using the quality metrics, Peak Signal to Noise Ratio (PSNR) (Sankur et al., 2002), the Structural Similarity Index (SSIM), and the Feature Similarity Index (FSIM) that were described in Table 1.1 in Chapter 1. The segmented images are presented in Figure 6.4.

Visually, the most interesting case is the gral_05 image with two thresholds, as the approach has removed most of the details of the sand, leaving the main shape of the snake and its shadow. Please refer to the segmented images of other chapters to highlight the difference. Since the shape of the image gral_05 is unimodal and symmetric, the algorithm seems to be struggling to place thresholds values over that simple histogram. The Masi formulation has a parameter r that can modify the overall result of the fitness function producing images that vary visually over the number of thresholds in comparison to other entropies where the segmentation is more consistent as the number of thresholds increases. In general, we can observe that each image's level of detail is consistent with the number of thresholds.

Regarding the quality of the outputs, Table 6.1 presents the results generated by the output images in terms of the PSNR, SSIM, and FSIM by using the PO in combination with Otsu's fitness function.

Table 6.1 presents the mean and the standard deviation (std) of the PSNR, SSIM, and FSIM values obtained by comparing the original grayscale image with the segmented image obtained by the thresholds computed using the EO. As in the previous

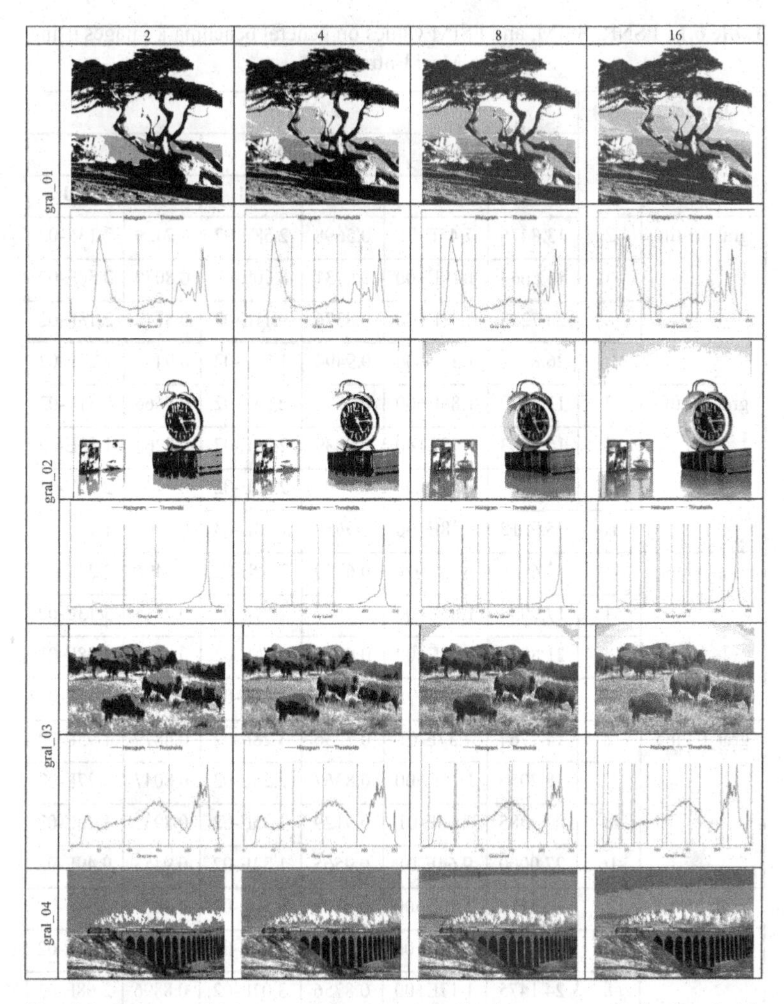

Fig. 6.4. Segmented images and the corresponding histogram with the thresholds obtained by using the EO with the Masi formulation of entropy.

Table 6.1. PSNR, SSIM, and FSIM values on general benchmark images using Masi Entropy.

Image	nTh	EO Masi					
		PSNR		SSIM		FSIM	
		mean	std	mean	std	mean	std
gral_01.tiff	2	13.8476	7.45E-01	0.5698	2.58E-02	0.7439	2.13E-02
	4	16.5663	1.19E+00	0.6931	8.70E-02	0.8070	2.50E-02
	8	20.7225	2.31E+00	0.8376	9.11E-02	0.8748	2.18E-02
	16	26.8165	1.34E+00	0.9494	1.72E-02	0.9436	1.25E-02
gral_02.tiff	2	11.8970	1.86E+00	0.7778	2.85E-02	0.7666	2.31E-02
	4	15.5083	1.98E+00	0.8839	2.25E-02	0.8263	1.83E-02
	8	21.1369	1.54E+00	0.9395	2.40E-02	0.8746	2.85E-02
	16	25.9212	1.38E+00	0.9659	7.13E-03	0.9204	1.17E-02
gral_03.jpg	2	13.6949	6.02E-01	0.6239	2.32E-02	0.6496	2.24E-02
	4	17.9033	1.04E+00	0.7777	2.71E-02	0.7814	3.13E-02
	8	21.8865	1.28E+00	0.8776	2.69E-02	0.8726	2.98E-02
	16	27.3145	7.68E-01	0.9523	1.14E-02	0.9472	9.16E-03
gral_04.jpg	2	14.1764	7.37E-01	0.7236	3.28E-02	0.6879	1.95E-02
	4	18.2056	1.25E+00	0.8367	2.25E-02	0.8047	3.27E-02
	8	22.4585	7.66E-01	0.9129	1.35E-02	0.8932	1.28E-02
	16	27.0000	9.64E-01	0.9565	1.34E-02	0.9453	9.49E-03
gral_05.tiff	2	14.2120	2.50E+00	0.6403	6.50E-02	0.4875	7.53E-02
	4	19.7043	2.37E+00	0.7369	8.43E-02	0.7486	6.61E-02
	8	24.1475	1.17E+00	0.8756	3.07E-02	0.8896	2.68E-02
	16	28.0930	1.23E+00	0.9416	1.74E-02	0.9536	1.51E-02

chapters, the results in the table are computed from a set of 35 independent experiments. The std permits one to see the stability of the algorithm, in lower dimension as in $th = 2$ the std values are closer to 0 and it means that the PO is more stable with the results. Meanwhile, for $th = 16$ the EO presents some fluctuations that are reflected in higher std. Regarding the mean values of the quality metrics, for all of them, when the number of thresholds increases, the values also increase. For PSNR, SSIM, and FSIM, it represents a higher similarity between the two images. In specific for SSIM and FSIM, when its values are closer to 1, the internal structure and features of the pixels from the segmented image are closer to the original image.

Following the discussion we started over the image gral_05, the numerical results show that the segmented image is competitive. We can also observe that the std is larger than other nonparametric formulations of the thresholding problem, as the Masi entropy includes the tuning parameter. These results indicate that the Masi formulation is powerful but not as stable as other approaches, leaving an attractive open area for research.

6.5 Conclusions

Most of the nonparametric criteria analyzed in this book are entropic measures with subtle differences that generate different results. From all of those, the formulation proposed by Masi is the most recent. It has excellent potential as it can behave like Rényi's entropy or Tsallis by just changing the value of the parameter r. This property can be exploited to create better thresholding approaches. The EO is a recently proposed methodology with a great deal of acceptance over different research communities. In the case of image segmentation, this method is a suitable alternative since it is easy to adapt any nonparametric criterion, in this case, Masi entropy.

Exercises

6.1 Implement the Masi entropy formulation for a single threshold presented in Eq. 6.12 in any programming language.

6.2 Use the exhaustive search to find the best threshold of the cameraman image using the cross-entropy for a single threshold.

6.3 Program the Masi entropy for multilevel thresholding presented in Eq. 6.12.

6.4 Implement the EO in combination with the cross-entropy for multilevel thresholding and test it over different images.

6.5 Program the PSNR, SSIM, and FSIM and verify the quality of the segmentation.

6.6 Implement the program created in Exercise 6.4 for color images.

References

Abdel-Basset, M., Mohamed, R., Mirjalili, S., Chakrabortty, R.K. and Ryan, M.J. (2020). Solar photovoltaic parameter estimation using an improved equilibrium optimizer. *Solar Energy*, 209: 694–708. https://doi.org/10.1016/j.solener.2020.09.032.

Abdul-hamied, D.T., Shaheen, A.M., Salem, W.A., Gabr, W.I. and El-sehiemy, R.A. (2020). Equilibrium optimizer based multi-dimensions operation of hybrid AC/DC grids. *Alexandria Engineering Journal*. https://doi.org/10.1016/j.aej.2020.08.043.

Dinh, P.H. (2021). Multimodal medical image fusion based on equilibrium optimizer algorithm and local energy functions. *Applied Intelligence*, 51(11): 8416–8431. https://doi.org/10.1007/S10489-021-02282-W/FIGURES/14.

Dinkar, S.K., Deep, K., Mirjalili, S. and Thapliyal, S. (2021). Opposition-based Laplacian Equilibrium Optimizer with application in Image Segmentation using Multilevel Thresholding. *Expert Systems with Applications*, 174: 114766. https://doi.org/10.1016/J.ESWA.2021.114766.

Faramarzi, A., Heidarinejad, M., Stephens, B. and Mirjalili, S. (2020). Equilibrium optimizer: A novel optimization algorithm. *Knowledge-Based Systems*, 191: 105190. https://doi.org/10.1016/j.knosys.2019.105190.

Gao, Y., Zhou, Y. and Luo, Q. (2020). An efficient binary equilibrium optimizer algorithm for feature selection. *IEEE Access*, 8: 140936–140963. https://doi.org/10.1109/ACCESS.2020.3013617.

Gui, P., He, F., Ling, B.W.K. and Zhang, D. (2021). United equilibrium optimizer for solving multimodal image registration. *Knowledge-based Systems*, 233: 107552. https://doi.org/10.1016/J.KNOSYS.2021.107552.

Kapur, J.N.N., Sahoo, P.K.K. and Wong, A.K.C.K.C. (1985). A new method for gray-level picture thresholding using the entropy of the histogram. *Computer Vision, Graphics, and Image Processing*, 29(3): 273–285. https://doi.org/10.1016/0734-189X(85)90125-2.

Lan, P., Xia, K., Pan, Y. and Fan, S. (2021). An improved equilibrium optimizer algorithm and its application in LSTM neural network. *Symmetry 2021*, 13(9): 1706. https://doi.org/10.3390/SYM13091706.

Li, Y., Fan, X. and Li, G. (2007). Image segmentation based on Tsallis-entropy and Rényi-entropy and their comparison. *2006 IEEE International Conference on Industrial Informatics, INDIN'06*, 00(i): 943–948. https://doi.org/10.1109/INDIN.2006.275704.

Mahmoudi, L. and El Zaart, A. (2012). A survey of entropy image thresholding techniques. *2012 2nd International Conference on Advances in Computational Tools for Engineering Applications,* ACTEA 2012: 204–209. https://doi.org/10.1109/ICTEA.2012.6462867.

Masi, M. (2005). A step beyond Tsallis and Rényi entropies. *Physics Letters, Section A*: *General, Atomic and Solid State Physics*, 338(3–5): 217–224. https://doi.org/10.1016/j.physleta.2005.01.094.

Naik, M.K., Panda, R. and Abraham, A. (2021). An opposition equilibrium optimizer for context-sensitive entropy dependency based multilevel thresholding of remote sensing images. *Swarm and Evolutionary Computation*, 65: 100907. https://doi.org/10.1016/J.SWEVO.2021.100907.

Sankur, B., Sankur, B. and Sayood, K. (2002). Statistical evaluation of image quality measures. *Journal of Electronic Imaging*, 11(2): 206. https://doi.org/10.1117/1.1455011.

Seleem, S.I., Hasanien, H.M. and El-Fergany, A.A. (2021). Equilibrium optimizer for parameter extraction of a fuel cell dynamic model. *Renewable Energy*, 169: 117–128. https://doi.org/10.1016/J.RENENE.2020.12.131.

Tang, A.D., Han, T., Zhou, H. and Xie, L. (2021). An improved equilibrium optimizer with application in unmanned aerial vehicle path planning. *Sensors,* 21(5): 1814. https://doi.org/10.3390/S21051814.

Too, J. and Mirjalili, S. (2020). General Learning Equilibrium Optimizer: A New Feature Selection Method for Biological Data Classification. Griffith University. *Https://Doi.Org/10.1080/08839514.2020.1861407*, 35(3): 247–263. https://doi.org/10.1080/08839514.2020.1861407.

Tsallis, C. (1988). Possible generalization of Boltzmann-Gibbs statistics. *Journal of Statistical Physics*, 52(1): 479–487. https://doi.org/10.1007/BF01016429.

Wunnava, A., Naik, M.K., Panda, R., Jena, B. and Abraham, A. (2020). A novel interdependence based multilevel thresholding technique using adaptive equilibrium optimizer. *Engineering Applications of Artificial Intelligence*, 94: 103836. https://doi.org/10.1016/j.engappai.2020.103836.

CHAPTER 7
MATLAB® Codes

This chapter is devoted to showing the code used in the algorithms, and it is analyzed and explained briefly for the enthusiastic reader. The structure given to the programs is organized in a series of files and follows the code's aim which is to be able to use them in a general way.

Keeping this in mind, it is possible to quickly adjust and implement any problem, knowing only the objective function and applying the changes in some conditionals. The codification was developed in MATLAB®, and each file has the extension .m.

7.1 Equilibrium Optimizer (EO)

We remember the analogy of the EO which is a novel optimization algorithm inspired by control volume mass balance and is to estimate the dynamic and equilibrium states. Therefore, it involves solutions of each particle with its concentration (position) to find the best equilibrium. The following function EO is where the implementation is developed to find the best solution. In a detailed view, the following features can be seen.

The first code is the main, where the parameters are set. It is widely used. Thirty runs make a pool of data, and after driving analysis, it shows the advantages and weaknesses. Between lines 13 to 15 is established the number of runs, number of particles used in the experiment, and the number of experiments. In line 19, the boundaries, dimension, and function objective is given; in this form, it is possible to use various functions to test the algorithm EO.

```
1  % main.m
2  % ---------------------------------------------------
3  % fobj = @YourCostFunction
4  % dim = number of your variables
5  % Max_iteration = maximum number of iterations
6  % Particles_no = number of particles (search agents)
7  % lb=[lb1,lb2,...,lbn] where lbn is the lower bound
   of variable n
8  % ub=[ub1,ub2,...,ubn] where ubn is the upper bound
   of variable n
9  % ---------------------------------------------------
10
11 clear all
12 clc
13 tic;
14 Run_no=30; % Number of independent runs
15 Particles_no=30; % Number of particles
16 Max_iteration=500; % Maximum number of iterations
17
18 Function_name='F1';
19
20 [lb,ub,dim,fobj]=Get_Functions_details(Function_
   name);
21
22 [Convergence_curve,Ave,Sd]=EO(Particles_no,Max_
   iteration,lb,ub,dim,fobj,Run_no);
23
24 display(['The average objective function is : ',
   num2str(Ave,7)]);
25 display(['The standard deviation is : ',
   num2str(Sd,7)]);
26
27 toc;
```

Between lines 5 to 8 are defined the particles named as equilibrium candidates; in 10, a vector of candidates is evaluated to find the best solutions. The algorithm will be controlled by the parameters: a1 to control the exploration, and a2 is a similar parameter for the exploitation. GP is the generation probability control of concentration updating by the generation. Lines 28 to 37 are devoted to improving the candidates and will be updated based on the new fitness shown in lines 41 to 52.

On the other side, it has been codifying the stage of exploitation. The aim is to use the same resources previously acquired (lines 58 to 71) using the equations developed in the algorithm, increasing the iteration and saving convergence curve and fitness. In the end, the data is usually presented and, on occasions, is reserved for further analysis and comparison.

```
1    function [Convergence_curve,Ave,Sd]=EO(Particles_
     no,Max_iter,lb,ub,dim,fobj,Run_no)
2
3    for irun=1:Run_no
4
5    Ceq1=zeros(1,dim); Ceq1_fit=inf;
6    Ceq2=zeros(1,dim); Ceq2_fit=inf;
7    Ceq3=zeros(1,dim); Ceq3_fit=inf;
8    Ceq4=zeros(1,dim); Ceq4_fit=inf;
9
10   C=initialization(Particles_no,dim,ub,lb);
11
12   Iter=0; V=1;
13
14   a1=2;
15   a2=1;
16   GP=0.5;
17
18   while Iter<Max_iter
19
20       for i=1:size(C,1)
21
22         Flag4ub=C(i,:)>ub;
```

```
23        Flag4lb=C(i,:)<lb;
24        C(i,:)=(C(i,:).*(~(Flag4ub+Flag4lb)))+ub.
          *Flag4ub+ lb.*Flag4lb;
25
26        fitness(i)=fobj(C(i,:));
27
28        if fitness(i)<Ceq1_fit
29          Ceq1_fit=fitness(i); Ceq1=C(i,:);
30        elseif fitness(i)>Ceq1_fit && fitness(i)<Ceq2_
          fit
31          Ceq2_fit=fitness(i); Ceq2=C(i,:);
32        elseif fitness(i)>Ceq1_fit && fitness(i)>Ceq2_
          fit && fitness(i)<Ceq3_fit
33           Ceq3_fit=fitness(i); Ceq3=C(i,:);
34        elseif fitness(i)>Ceq1_fit && fitness(i)>Ceq2_
          fit &&     fitness(i)>Ceq3_fit              &&
          fitness(i)<Ceq4_fit
35           Ceq4_fit=fitness(i); Ceq4=C(i,:);
36
37        end %end of if
38     end %end of for
39
40  %---------------- Memory saving-------------------
41     if Iter==0
42          fit_old=fitness;   C_old=C;
43     end
44
45        for i=1:Particles_no
46            if fit_old(i)<fitness(i)
47            fitness(i)=fit_old(i); C(i,:)=C_old(i,:);
48            end
49        end
50
51  C_old=C;
52  fit_old=fitness;
53  %-------------------------------------------------
54
55  Ceq_ave=(Ceq1+Ceq2+Ceq3+Ceq4)/4; % averaged candidate
56  C_pool=[Ceq1; Ceq2; Ceq3; Ceq4; Ceq_ave];%   Equilibrium
    pool
57
```

```
58  t=(1-Iter/Max_iter)^(a2*Iter/Max_iter);
    % Eq (9)
59
50
61      for i=1:Particles_no
62          lambda=rand(1,dim);    % lambda in Eq(11)
63          r=rand(1,dim); % r in Eq(11)
64          Ceq=C_pool(randi(size(C_pool,1)),:);     % random
            selection of one
                                 % candidate from the pool
65          F=a1*sign(r-0.5).*(exp(-lambda.*t)-1);   % Eq(11)
66          r1=rand(); r2=rand(); % r1 and r2 in Eq(15)
67          GCP=0.5*r1*ones(1,dim)*(r2>=GP);% Eq(15)
68          G0=GCP.*(Ceq-lambda.*C(i,:)); % Eq(14)
69          G=G0.*F; % Eq(13)
70          C(i,:)=Ceq+(C(i,:)-Ceq).*F+(G./
            lambda*V).*(1-F); % Eq(16)
71      end
72
73        Iter=Iter+1;
74        Convergence_curve(Iter)=Ceq1_fit;
75        Ceqfit_run(irun)=Ceq1_fit;
76
77  end   %end of the while
78
79
80     display(['Run no : ', num2str(irun)]);
81     display(['The best solution obtained by EO is :
       ', num2str(Ceq1,10)]);
82     display(['The best optimal value of the objective
       funciton found by EO is : ',
       num2str(Ceq1_fit,10)]);
83     disp(sprintf('--------------------------------------'));
84  end %end of the for, it is the current run the
    algorithm.
85
86  Ave=mean(Ceqfit_run);
87  Sd=std(Ceqfit_run);
88  end  %end of the function
```

As seen in the function EO, in line 10, the function initially is used to create the first pool of data, there the candidates are defined based on the dimension numbers and the upper and lower limits of their boundaries.

```
1   % This function initialize the first population of
    particles
2   function [Cin,domain]=initialization(SearchAgents_
    no,dim,ub,lb)
3
4   Boundary_no= size(ub,2); % number of boundaries
5
6   % If the boundaries of all variables are equal and
    user enter a single
7   % number for both ub and lb
8   if Boundary_no==1
9        Cin=rand(SearchAgents_no,dim).*(ub-lb)+lb;
10       domain=ones(1,dim)*(ub-lb);
11  end
12
13
14  % If each variable has a different lb and ub
15  if Boundary_no>1
16       for i=1:dim
17            ub_i=ub(i);
18            lb_i=lb(i);
19              Cin(:,i)=rand(SearchAgents_no,1).*(ub_i-
    lb_i)+lb_i;
20       end
21       domain=ones(1,dim).*(ub-lb);
22  end
```

Get_functions_details is an important function which is also is explainable. The search space is bounded by the parameters given as box constraints. Thus, the minimum and maximum value of each design variable should be provided. The dim defines the number of dimensions; in other words, the number of variables involved in the problem altogether provides the fitness, but all parameters are used directly in the objective function.

```
1    % lb is the lower bound: lb=[lb_1,lb_2,...,lb_d]
2    % up is the uppper bound: ub=[ub_1,ub_2,...,ub_d]
3    % dim is the number of variables (dimension of the
     problem)
4
5    function   [lb,ub,dim,fobj]   =   Get_Functions_
     details(F)
6
7
8    switch F
9        case 'F1'
10           fobj = @F1;
11           lb=-100;
12           ub=100;
13           dim=30;
14
15       case 'F2'
16           fobj = @F2;
17           lb=-100;
18           ub=100;
19           dim=30;
20
21       case 'F3'
22           fobj = @F3;
23           lb=-100;
24           ub=100;
25           dim=30;
26
27       case 'F4'
28           fobj = @F4;
29           lb=-100;
30           ub=100;
31           dim=30;
32
33       case 'F5'
34           fobj = @F5;
35           lb=-30;
36           ub=30;
37           dim=30;
38
```

```
39      case 'F6'
40          fobj = @F6;
41          lb=-100;
42          ub=100;
43          dim=30;
44
45      case 'F7'
46          fobj = @F7;
47          lb=-1.28;
48          ub=1.28;
49          dim=30;
50
51      case 'F8'
52          fobj = @F8;
53          lb=-500;
54          ub=500;
55          dim=30;
56
57      case 'F9'
58          fobj = @F9;
59          lb=-5.12;
50          ub=5.12;
61          dim=30;
62
63      case 'F10'
64          fobj = @F10;
65          lb=-32;
66          ub=32;
67          dim=30;
68
69      case 'F11'
70          fobj = @F11;
71          lb=-600;
72          ub=600;
73          dim=30;
74
75      case 'F12'
76          fobj = @F12;
77          lb=-50;
78          ub=50;
```

```
79              dim=30;
80
81          case 'F13'
82              fobj = @F13;
83              lb=-50;
84              ub=50;
85              dim=30;
86
87          case 'F14'
88              fobj = @F14;
89              lb=-65.536;
90              ub=65.536;
91              dim=2;
92
93          case 'F15'
94              fobj = @F15;
95              lb=-5;
96              ub=5;
97              dim=4;
98
99
100         case 'F16'
101             fobj = @F16;
102             lb=-5;
103             ub=5;
104             dim=2;
105
106         case 'F17'
107             fobj = @F17;
108             lb=[-5,0];
109             ub=[10,15];
110             dim=2;
111
112         case 'F18'
113             fobj = @F18;
114             lb=-2;
115             ub=2;
116             dim=2;
117
```

```
118         case 'F19'
119             fobj = @F19;
120             lb=0;
121             ub=1;
122             dim=3;
123
124         case 'F20'
125             fobj = @F20;
126             lb=0;
127             ub=1;
128             dim=6;
129
130         case 'F21'
131             fobj = @F21;
132             lb=0;
133             ub=10;
134             dim=4;
135
136         case 'F22'
137             fobj = @F22;
138             lb=0;
139             ub=10;
140             dim=4;
141
142         case 'F23'
143             fobj = @F23;
144             lb=0;
145             ub=10;
146             dim=4;
147
148
149     end
150
151     end
152
153     % F1
154
155     function o = F1(x)
156     o=sum(x.^2);
157     end
```

```
158
159 % F2
150
161 function o = F2(x)
162 o=sum(abs(x))+prod(abs(x));
163 end
164
165 % F3
166
167 function o = F3(x)
168 dim=size(x,2);
169 o=0;
170 for i=1:dim
171     o=o+sum(x(1:i))^2;
172 end
173 end
174
175 % F4
176
177 function o = F4(x)
178 o=max(abs(x));
179 end
180
181 % F5
182
183 function o = F5(x)
184 dim=size(x,2);
185 o=sum(100*(x(2:dim)-(x(1:dim-
    1).^2)).^2+(x(1:dim-1)-1).^2);
186 end
187
188 % F6
189
190 function o = F6(x)
191 o=sum(abs((x+.5)).^2);
192 end
193
194 % F7
195
196 function o = F7(x)
```

```
197  dim=size(x,2);
198  o=sum([1:dim].*(x.^4))+rand;
199  end
200
201  % F8
202
203  function o = F8(x)
204  o=sum(-x.*sin(sqrt(abs(x))));
205  end
206
207  % F9
208
209  function o = F9(x)
210  dim=size(x,2);
211  o=sum(x.^2-10*cos(2*pi.*x))+10*dim;
212  end
213
214  % F10
215
216  function o = F10(x)
217  dim=size(x,2);
218  o=-20*exp(-.2*sqrt(sum(x.^2)/dim))-
     exp(sum(cos(2*pi.*x))/dim)+20+exp(1);
219  end
220
221  % F11
222
223  function o = F11(x)
224  dim=size(x,2);
225  o=sum(x.^2)/4000-prod(cos(x./sqrt([1:dim])))+1;
226  end
227
228  % F12
229
230  function o = F12(x)
231  dim=size(x,2);
232  o=(pi/dim)*(10*((sin(pi*(1+(x(1)+1)/4)))^2)+
     sum((((x(1:dim-1)+1)./4).^2).*...
233  (1+10.*((sin(pi.*(1+(x(2:dim)+1)./4)))).^2))+
     ((x(dim)+1)/4)^2)+sum(Ufun(x,10,100,4));
```

```
234 end
235
236
237 % F13
238
239 function o = F13(x)
240 dim=size(x,2);
241 o=.1*((sin(3*pi*x(1)))^2+sum((x(1:dim-1)-
    1).^2.*(1+(sin(3.*pi.*x(2:dim))).^2))+...
242 ((x(dim)-1)^2)*(1+(sin(2*pi*x(dim)))^2))+
    sum(Ufun(x,5,100,4));
243 end
244
245 % F14
246
247 function o = F14(x)
248 aS=[-32 -16 0 16 32 -32 -16 0 16 32 -32 -16 0 16
    32 -32 -16 0 16 32 -32 -16 0 16 32;,...
249 -32 -32 -32 -32 -32 -16 -16 -16 -16 -16 0 0 0 0 0
    16 16 16 16 16 32 32 32 32 32];
250
251 for j=1:25
252     bS(j)=sum((x'-aS(:,j)).^6);
253 end
254 o=(1/500+sum(1./([1:25]+bS))).^(-1);
255 end
256
257 % F15
258
259 function o = F15(x)
250 aK=[.1957 .1947 .1735 .16 .0844 .0627 .0456 .0342
    .0323 .0235 .0246];
261 bK=[.25 .5 1 2 4 6 8 10 12 14 16];bK=1./bK;
262 o=sum((aK-((x(1).*(bK.^2+x(2).*bK))./
    (bK.^2+x(3).*bK+x(4)))).^2);
263 end
264
265 % F16
266
267 function o = F16(x)
```

```
268 o=4*(x(1)^2)-2.1*(x(1)^4)+(x(1)^6)/3+x(1)*x(2)-
    4*(x(2)^2)+4*(x(2)^4);
269 end
270
271 % F17
272
273 function o = F17(x)
274 o=(x(2)-(x(1)^2)*5.1/(4*(pi^2))+5/pi*x(1)-
    6)^2+10*(1-1/(8*pi))*cos(x(1))+10;
275 end
276
277 % F18
278
279 function o = F18(x)
280 o=(1+(x(1)+x(2)+1)^2*(19-14*x(1)+3*(x(1)^2)-
    14*x(2)+6*x(1)*x(2)+3*x(2)^2))*...
281   ( 3 0 + ( 2 * x ( 1 ) - 3 * x ( 2 ) ) ^ 2 * ( 1 8 -
      3 2 * x ( 1 ) + 1 2 * ( x ( 1 ) ^ 2 ) + 4 8 * x ( 2 ) -
      36*x(1)*x(2)+27*(x(2)^2)));
282 end
283
284 % F19
289
286 function o = F19(x)
287 aH=[3 10 30;.1 10 35;3 10 30;.1 10 35];cH=[1 1.2
    3 3.2];
288 pH=[.3689 .117 .2673;.4699 .4387 .747;.1091 .8732
    .5547;.03815 .5743 .8828];
289 o=0;
290 for i=1:4
291  o=o-cH(i)*exp(-(sum(aH(i,:).*((x-pH(i,:)).^2))));
292 end
293 end
294
295 % F20
296
297 function o = F20(x)
298 aH=[10 3 17 3.5 1.7 8;.05 10 17 .1 8 14;3 3.5 1.7
    10 17 8;17 8 .05 10 .1 14];
299 cH=[1 1.2 3 3.2];
```

```
301 pH=[.1312  .1696  .5569  .0124  .8283  .5886;.2329
    .4135 .8307 .3736 .1004 .9991;...
302 .2348 .1415 .3522 .2883 .3047 .6650;.4047 .8828
    .8732 .5743 .1091 .0381];
303 o=0;
304 for i=1:4
305  o=o-cH(i)*exp(-(sum(aH(i,:).*((x-pH(i,:)).^2))));
306 end
307 end
308
309 % F21
310
311 function o = F21(x)
312 aSH=[4 4 4 4;1 1 1 1;8 8 8 8;6 6 6 6;3 7 3 7;2 9
    2 9;5 5 3 3;8 1 8 1;6 2 6 2;7 3.6 7 3.6];
313 cSH=[.1 .2 .2 .4 .4 .6 .3 .7 .5 .5];
314
315 o=0;
316 for i=1:5
317  o=o-((x-aSH(i,:))*(x-aSH(i,:))'+cSH(i))^(-1);
318 end
319 end
320
321 % F22
322
323 function o = F22(x)
324 aSH=[4 4 4 4;1 1 1 1;8 8 8 8;6 6 6 6;3 7 3 7;2 9
    2 9;5 5 3 3;8 1 8 1;6 2 6 2;7 3.6 7 3.6];
325 cSH=[.1 .2 .2 .4 .4 .6 .3 .7 .5 .5];
326
327 o=0;
328 for i=1:7
329  o=o-((x-aSH(i,:))*(x-aSH(i,:))'+cSH(i))^(-1);
330 end
331 end
332
333 % F23
334
335 function o = F23(x)
```

```
336 aSH=[4 4 4 4;1 1 1 1;8 8 8 8;6 6 6 6;3 7 3 7;2 9
    2 9;5 5 3 3;8 1 8 1;6 2 6 2;7 3.6 7 3.6];
337 cSH=[.1 .2 .2 .4 .4 .6 .3 .7 .5 .5];
338
339 o=0;
340 for i=1:10
341  o=o-((x-aSH(i,:))*(x-aSH(i,:))'+cSH(i))^(-1);
342 end
343 end
344
345 function o=Ufun(x,a,k,m)
346    o=k.*((x-a).^m).*(x>a)+k.*((-x-a).^m).*(x<(-a));
347 end
```

7.2 Political Optimizer (PO)

The multiphase process of politics inspired the Political Optimizer. In particular, society has its own rules to have candidates, parties, and parliaments. Reflecting the diversity and coexistence to have the best candidates, this is the analogy of the best men for the job in the algorithm. Contemplating on this, the following codes were developed: government formation, election campaign, party switching, parliamentarian.

The main file manages the portfolio and calls the function PO. It is easy to see the use, and adjustment of the parameters before running the algorithm in lines 3 to 5 establish the number of parties and the quantity of interchanges of candidates. Line 15 is called the function to get details of each test, and it could be interchanging this parameter to adapt the values in the problems of optimization. The objective function is called in line 21 to process the setting data and return three values: best score, best position, and the fitness curve in the time of use. In this case, lines 28 to 32 calculate the statistics to print them later.

```
1   % main.m
2   %     Adjustable parameters
3   parties = 8;          %Number of political parties
4   lambda = 1.0;         %Max limit of party switching
    rate
5   fEvals = 30000;     %Number of function evaluations
6   %%%%%%%%%%%%%%%%%%%%%%%%%%%%%%%%%%%%%%%%%%%%%%%%%
7   areas = parties;
8   populationSize=parties * areas; % Number of search
    agents
9   Max_iteration = round(fEvals / (parties * areas +
    areas));
10  runs = 10;
11
12  for fn = 1:23
13
14    Function_name=strcat('F',num2str(fn)); % Name of
      the test function
15    [lb,ub,dim,fobj]=Get_Functions_
      details(Function_name);
16
17       % Calling algorithm
18        Best_score_T = zeros(1,runs);
19        for run=1:runs
20        rng('shuffle');
21       [Best_score_0,Best_pos,PO_cg_curve]=
         PO(populationSize,areas,parties,lambda,
         Max_iteration,lb,ub,dim,fobj);
22       Best_score_T(1,run) = Best_score_0;
23
24       Best_score_0
25      end
26
27   %Finding statistics
28   Best_score_Best = min(Best_score_T);
29   Best_score_Worst = max(Best_score_T);
30   Best_score_Median = median(Best_score_T,2);
31   Best_Score_Mean = mean(Best_score_T,2);
32   Best_Score_std = std(Best_score_T);
33
```

```
34   %Printing results
35   display(['Fn = ', num2str(fn)]);
36   display(['Best, Worst, Median, Mean, and Std. are
     as: ',num2str(Best_score_Best),'  ', ...
37       num2str(Best_score_Worst),'  ', num2str(Best_
         score_Median),'  ',
         num2str(Best_Score_Mean),'  ', num2str(Best_
         Score_std)]);
38
39  end  %end of the iterator to recover all functions
    to test
```

As we saw previously, the use of this function receives the following inputs: SearchAgentes, areas of the parties, iterations, and the parameters of the problems, including the boundaries' upper and lower limits, the dimensions, and the objective function. Between lines 6 to 11 is a process to initialize the values of agents and the auxiliary variables. In this case, the PO function makes use of procedures to compact and show a clean code. In line 17, run the Government Formation, and the main loop contains the stages of the analogy between lines 26 to 30. These procedures will be discussed below.

```
1   function   [Leader_score,Leader_pos,Convergence_
    curve]=...
    PO(SearchAgents_no,areas,parties,lambda,Max_
    iter,lb,ub,dim,fobj)
2   % initialize position vector and score for the
    leader
3   Leader_pos=zeros(1,dim);
4   Leader_score=inf; %change this to -inf for
    maximization problems
5
6   %Initialize the positions of search agents
7   Positions=initialization(SearchAgents_
    no,dim,ub,lb);
8   auxPositions = Positions;
9   prevPositions = Positions;
10  Convergence_curve=zeros(1,Max_iter);
```

```
11  fitness=zeros(SearchAgents_no, 1);
12
13  %Running phases for initializations
14  Election;   %Run election phase
15  auxFitness = fitness;
16  prevFitness = fitness;
17  GovernmentFormation;
18
19  t=0;% Loop counter
20           while t<Max_iter
21           prevFitness = auxFitness;
22           prevPositions = auxPositions;
23           auxFitness = fitness;
24           auxPositions = Positions;
25
26           ElectionCampaign;
27           PartySwitching;
28           Election;
29           GovernmentFormation;
30           Parliamentarism;
31
32           t=t+1;
33           Convergence_curve(t)=Leader_score;
34           [t Leader_score];
35           end
36  end %end of the function
```

The function of initialization returns each search agent's position values based on the dimension of the problem, with the limitations of the boundaries.

```
1   % This function initialize the first population of
    search agents
2   function   Positions=initialization(SearchAgents_
    no,dim,ub,lb)
3   Boundary_no= size(ub,2); % numnber of boundaries,
    bounds are vector
4   Positions       =       zeros(SearchAgents_no,dim);
    %Declaration
5
```

```
6   % If each variable has a different lb and ub
7      if Boundary_no>1
8          for i=1:dim
9              ub_i=ub(i);
10             lb_i=lb(i);
11             Positions(:,i)=rand(SearchAgents_
               no,1).*(ub_i-lb_i)+lb_i;
12         end
13     end
14
15  % If the boundaries of all variables are equal and
    user enter a single
16  % number for both ub and lb
17     if Boundary_no==1
18     Positions=rand(SearchAgents_no,dim).*(ub-lb)+lb;
19     end
20  end
```

The election campaign is divided into two large evaluation groups: one is for the values greater than fitness values (lines 16 to 32), and the other is for the lower cases (lines 36 to 53). In turn, they are divided into three similar cases. The three cases represent the exploitation near the position based in the radio and center of the previous position and the actual position. In the analogy represent, the candidates are on the campaign trying to get the best popularity (new fitness) to be the best election contender to the parties.

```
1  % Election campaign
2  for whichMethod = 1:2
3   for a = 1:areas
4    for p = 1:parties
5     i = (p-1)*areas + a; %index of member
6
7     for j=1:dim
8       if whichMethod == 1    position-updating w.r.t
   party leader
9        center = pLeaders(p,j);
```

```
10    elseif whichMethod == 2%position-updating w.r.t
   area winner
11      center = aWinners(a,j);
12     end
13
14 %Cases of Eq. 9 in paper
15   if prevFitness(i) >= fitness(i)
16     if  (prevPositions(i,j)  <=  Positions(i,j)  &&
       Positions(i,j) <= center) ...
17       ||  (prevPositions(i,j)  >=  Positions(i,j)  &&
        Positions(i,j) >= center)
18
19      radius = center - Positions(i,j);
20      Positions(i,j) = center + rand() * radius;
21    elseif (prevPositions(i,j) <= Positions(i,j) &&
      Positions(i,j) >= center ...
          && center >= prevPositions(i,j))
22        || (prevPositions(i,j) >= Positions(i,j) &&
          Positions(i,j) <= center ...
23        && center <= prevPositions(i,j))
24
25         radius = abs(Positions(i,j) - center);
26         Positions(i,j)  =  center  +  (2*rand()-1)  *
           radius;
27      elseif  (prevPositions(i,j)  <=  Positions(i,j)
        && Positions(i,j) >= center     &&  center  <=
        prevPositions(i,j)) ...
28        || (prevPositions(i,j) >= Positions(i,j) &&
          Positions(i,j) <= center &&          center
          >= prevPositions(i,j))
29
30        radius = abs(prevPositions(i,j) - center);
31        Positions(i,j)  =  center  +  (2*rand()-1)  *
          radius;
32       end
33
34                     %Cases of Eq. 10 in paper
35   elseif prevFitness(i) < fitness(i)
36       if (prevPositions(i,j) <= Positions(i,j) &&
   Positions(i,j) <= center) ...
```

```
37          || (prevPositions(i,j) >= Positions(i,j)
            && Positions(i,j) >= center)
38
39
40          radius = abs(Positions(i,j) - center);
41          Positions(i,j) = center + (2*rand()-1) *
            radius;
42        elseif (prevPositions(i,j) <= Positions(i,j)
          && Positions(i,j) >= center && center >=
          prevPositions(i,j)) ...
43            || (prevPositions(i,j) >= Positions(i,j)
            && Positions(i,j) <= center && center <=
            prevPositions(i,j))
44
45          radius = Positions(i,j) - prevPositions(i,j);
46          Positions(i,j) = prevPositions(i,j) + rand()
            * radius;
47        elseif (prevPositions(i,j) <= Positions(i,j)
        && Positions(i,j) >= center &&         center
        <= prevPositions(i,j)) ...
48              ||revPositions(i,j) >= Positions(i,j) &&
              Positions(i,j) <= center &&
              center >= prevPositions(i,j))
49
50                  center2 = prevPositions(i,j);
51                  radius = abs(center - center2);
52                  Positions(i,j) = center + (2*rand()-
                    1) * radius;
53          end
54    end
55
56      end
57    end
58   end
59  end
```

The party switching is the phase to balance the parties due to control parameters on line 4 lamda, which is analogous to creating a balance between exploitation and exploration. During the time, will no occurrence change between parties that retained the best candidates?

```
1   % Party switching Phase
2   psr = (1-t*((1)/Max_iter)) * lambda;
3
4   for p=1:parties
5       for a=1:areas
6           fromPInd = (p-1)*areas + a;
7           if rand() < psr
8               %Selecting a party other than current
                where to want to send the
9               %member
10              toParty = randi(parties);
11              while(toParty == p)
12                  toParty = randi(parties);
13              end
14
15              %Deciding member in topParty
16               toPStInd = (toParty-1) * areas + 1;
17               toPEndIndex = toPStInd + areas - 1;
18               [~,toPLeastFit] =
                max(fitness(toPStInd:toPEndIndex));
19               toPInd = toPStInd + toPLeastFit-1;
20
21
22              %Deciding what to do with member in
                FROM party and switching
23              fromPInd = (p-1)*areas + a;
24              temp = Positions(toPInd,:);
25              Positions(toPInd,:) =
                Positions(fromPInd);
26              Positions(fromPInd,:)=temp;
27
28              temp = fitness(toPInd);
29              fitness(toPInd) = fitness(fromPInd);
30              fitness(fromPInd) = temp;
31          end
32      end
33  end
```

Once completed, the adjustment into parties is coming to the process of election. In this phase, the parties select and interchange members. It is based on the position and quality, keeping the best candidates with higher fitness. On the other hand, the candidates less qualified are discarded. Previously analyzed and adjusted the candidate, if they have gone out of the bounds, it could be seen in lines 4 to 9.

```
1   % Election %%%%%%%%%%%%%%%%%%%%%%%%%%%%%%%%%%%%%%
2   for i=1:size(Positions,1)
3   % Return back the search agents that go beyond the
    boundaries of the search space
4      Flag4ub=Positions(i,:)>ub;
5      Flag4lb=Positions(i,:)<lb;
6      Positions(i,:)=(Positions(i,:).*
       (~(Flag4ub+Flag4lb)))...
7      +ub.*Flag4ub+lb.*Flag4lb;
8      %Calculate objective function for each search
       agent
9      fitness(i,1)=fobj(Positions(i,:));
10
11     %Update the leader
12   if fitness(i,1)<Leader_score % Change this to >
     for maximization problem
13    Leader_score=fitness(i,1);
14    Leader_pos=Positions(i,:);
15      end
16  end
```

The government formation is a phase where the main objective is to get the best candidate of each party; it could be seen in lines 4 to 8, and reserve to the phase of parliament; at same time it localizes the party leader.

```
1   % Government Formation
2   aWinnerInd=zeros(areas,1);      %Indices of area
    winners in x
3   aWinners = zeros(areas,dim); %Area winners are
    stored separately
4   for a = 1:areas
```

```
5      [aWinnerFitness,aWinnerParty]=       min(fitness
       (a:areas:SearchAgents_no));
6      aWinnerInd(a,1) = (aWinnerParty-1) * areas + a;
7     aWinners(a,:) = Positions(aWinnerInd(a,1),:);
8   end
9
10 %Finding party leaders
11 pLeaderInd=zeros(parties,1);      %Indices of party
   leaders in x
12 pLeaders = zeros(parties,dim);  %Positions of party
   leaders in x
13 for p = 1:parties
14      pStIndex = (p-1) * areas + 1;
15      pEndIndex = pStIndex + areas - 1;
16      [partyLeaderFitness,leadIndex]=
        min(fitness(pStIndex:pEndIndex));
17      pLeaderInd(p,1) = (pStIndex - 1) + leadIndex;
        %Indexof party leader
18     pLeaders(p,:) = Positions(pLeaderInd(p,1),:);
19 end
```

The phase of parliament is the beginning of analysis for the winner of each party and get modifications in the dimensions to get the best characteristic of the winners. It will form a new candidate in lines 11 to 14, and if this is better, the best candidate (line 15) is replaced in lines 17 to 22.

```
1  %Parliamentarism
2  for a=1:areas
3    newAWinner = aWinners(a,:);
4    i = aWinnerInd(a);
5
6    toa = randi(areas);
7    while(toa == a)
8          toa = randi(areas);
9    end
10         toAWinner = aWinners(toa,:);
11   for j = 1:dim
12     distance = abs(toAWinner(1,j) - newAWinner(1,j));
```

```
13      newAWinner(1,j) = toAWinner(1,j) + (2*rand()-1) *
        distance;
14     end
15     newAWFitness=fobj(newAWinner(1,:));
16
17      %Replace only if improves
18      if newAWFitness < fitness(i)
19              Positions(i,:) = newAWinner(1,:);
20              fitness(i) = newAWFitness;
21              aWinners(a,:) = newAWinner(1,:);
22      end
23 end
```

7.3 Manta Ray Foraging Optimization (MRFO)

As we see now, exist the different sources of inspiration, and is recurrent the nature and the animal behavior. The case of the MRFO is not exclusive, and therefore, it sees the algorithm structure and analysis of the coding as essential.

First, it could see the main file setting the values to run the MFRO function. In this case, line 6 establishes the function to work; as we can see in the other algorithms, this action takes place before the call of the function objective. It is an interesting variation because the pre-processing of the problem is done at the beginning of MRFO. It is seen in the following code. Lines 10 to 22 show the results for visual inspection, helping to understand the algorithm's behavior.

```
1   clc;
2   clear;
3
4   MaxIteration=1000;
5   PopSize=50;
6   FunIndex=5;
7   [BestX,BestF,HisBestF]=MRFO(FunIndex,
    MaxIteration,PopSize);
8
9
```

```
10 display(['F_index=', num2str(FunIndex)]);
11 display(['The best fitness is: ', num2str(BestF)]);
12 Optimal(FunIndex)=BestF;
13
14 %display(['The best solution is: ', num2str(BestX)]);
15  if BestF>=0
16      semilogy(HisBestF,'r','LineWidth',2);
17  else
18      plot(HisBestF,'r','LineWidth',2);
19  end
20  xlabel('Iterations');
21  ylabel('Fitness');
22  title(['F',num2str(FunIndex)]);
```

Following the MRFO function, we receive three parameters: F_index is the value to get the setting of the function; this is used in line 17 and sets the lower and upper boundary and the dimension of the problem. The initial population is created between lines 19 to 22; this does not mean they are the best. The best adjustments are made on lines 26 to 31. The algorithm is represented in three groups of coding, which represent the behavior of the manta ray. First, line 42 evaluates the spiral movement, changing the position of the manta ray. Second, lines 45 and 46 search for food-based into the same place. And third, lines 49 and 50 are the process of following the leader of the shoal. These evaluations are repeated in lines 54 to 66. To increase the search diversity, create a new position in lines 69 to 75 to avoid the food shortage. The update is made in lines 82 to 96 to improve the population of a manta ray.

```
1  % FunIndex: Index of function.
2  % MaxIt: The maximum number of iterations.
3  % PopSize: The size of population.
4  % PopPos: The position of population.
5  % PopFit: The fitness of population.
6  % Dim: The dimensionality of prloblem.
7  % Alpha: The weight coefficient in chain foraging.
8  % Beta: The weight coefficient in cyclone foraging.
9  % S: The somersault factor.
```

```
10 % BestF: The best fitness corresponding to BestX.
11 % HisBestFit: History best fitness over iterations.
12 % Low: The low bound of search space.
13 % Up: The up bound of search space.
14
15 function [BestX,BestF,HisBestFit]=MRFO
   (F_index,MaxIt,nPop)
16
17              [Low,Up,Dim]=FunRange(F_index);
18
19       for i=1:nPop
20           PopPos(i,:)=rand(1,Dim).*(Up-Low)+Low;
21           PopFit(i)=BenFunctions(PopPos(i,:),F_
             index,Dim);
22       end
23          BestF=inf;
24          BestX=[];
25
26       for i=1:nPop
27           if PopFit(i)<=BestF
28              BestF=PopFit(i);
29              BestX=PopPos(i,:);
30           end
31       end
32
33          HisBestFit=zeros(MaxIt,1);
34
35
36 for It=1:MaxIt
37        Coef=It/MaxIt;
38
39          if rand<0.5
40             r1=rand;
41         Beta=2*exp(r1*((MaxIt-It+1)/
           MaxIt))*(sin(2*pi*r1));
42         if  Coef>rand
43           newPopPos(1,:)=BestX+rand(1,Dim).*(BestX-
             PopPos(1,:))+Beta*(BestX-PopPos(1,:));%Equation
             (4)
44       else
```

```
45          IndivRand=rand(1,Dim).*(Up-Low)+Low;
46          newPopPos(1,:)=IndivRand+rand(1,Dim).
            *(IndivRand-    P o p P o s ( 1 , : ) ) +
            Beta*(IndivRand-PopPos(1,:)); %Equation (7)
47            end
48        else
49            Alpha=2*rand(1,Dim).*(-log(rand(1,D
            im))).^0.5;
50            newPopPos(1,:)=PopPos(1,:)+rand(1,Dim).*
            (BestX-PopPos(1,:))+Alpha.*(BestX-
            PopPos(1,:)); %Equation (1)
51        end
52
53    for i=2:nPop
54          if rand<0.5
55             r1=rand;
56             Beta=2*exp(r1*((MaxIt-It+1)/
               MaxIt))*(sin(2*pi*r1));
57                if  Coef>rand
58                   newPopPos(i,:)=BestX+rand(1,Dim).*
                     ( P o p P o s ( i - 1 , : ) -
                     PopPos(i,:))+Beta*(BestX-
                     PopPos(i,:)); %Equation (4)
59                else
60                  IndivRand=rand(1,Dim).*(Up-Low)+Low;
61                  newPopPos(i,:)=IndivRand+rand(1,
                    Dim).*(PopPos(i-1,:)-PopPos(i,:))
                    +Beta*(IndivRand-PopPos(i,:));
                    %Equation (7)
62                end
63          else
64            Alpha=2*rand(1,Dim).*(-log(rand(1,D
            im))).^0.5;
65            newPopPos(i,:)=PopPos(i,:)+rand(1,Dim).*
            (PopPos(i-1,:)-PopPos(i,:))+ Alpha.*
            (BestX-PopPos(i,:)); %Equation (1)
66        end
67    end
68
69             for i=1:nPop
```

```
70        newPopPos(i,:)=SpaceBound(newPopPos(i,:),
          Up,Low);newPopFit(i)=
          BenFunctions(newPopPos(i,:),F_index,Dim);
71                 if newPopFit(i)<PopFit(i)
72                    PopFit(i)=newPopFit(i);
73                    PopPos(i,:)=newPopPos(i,:);
74                 end
75              end
76
77               S=2;
78           for i=1:nPop
79            newPopPos(i,:)=PopPos(i,:)+S*(rand*BestX-
              rand* PopPos(i,:)); %Equation (8)
80           end
81
82        for i=1:nPop
83            newPopPos(i,:)=SpaceBound(newPopPos(i,:),
              Up,Low);
84            newPopFit(i)=BenFunctions(newPopPos(i,:),
              F_index,Dim);
85            if newPopFit(i)<PopFit(i)
86               PopFit(i)=newPopFit(i);
87               PopPos(i,:)=newPopPos(i,:);
88            end
89        end
90
91        for i=1:nPop
92           if PopFit(i)<BestF
93              BestF=PopFit(i);
94              BestX=PopPos(i,:);
95           end
96        end
97
98        HisBestFit(It)=BestF;
99 end
```

In general, we have seen strategies to keep the values of the individual within the boundaries, and this case is not the exception. The function space-bound adjusts the values to respect the lower and upper boundaries.

```
1 function  X=SpaceBound(X,Up,Low)
2
3       Dim=length(X);
4       S=(X>Up)+(X<Low);
5       X=(rand(1,Dim).*(Up-Low)+Low).*S+X.*(~S);
6
7 end
```

7.4 Archimedes Optimization Algorithm (AOA)

The AOA is based on the principle of Archimedes, it is physical law where the Greek Archimedes see the body's behavior when submerged in a fluid, and the buoyancy displaced the same volume in an opposed reaction. This is taken as inspiration for the next implementation. As characteristic in these developments, a function is implemented to pass the problem parameters and can be tested with different values, called main. Lines 4 to 11 set the values; in this sense, it could change in other problems respecting the structure of values.

```
1   clc;
2   clear all;
3   close all
4   fobj = @sumsqu;nvar=30;
5   lb=-10;
6   ub=10;
7   Materials_no=30;
8   Max_iter=1000;
9   dim=10;
10  % C3=2;C4=.5;       %cec and engineering problems
11  C3=1;C4=2;      %standard Optimization functions
12  [Xbest, Scorebest,Convergence_curve]=AOA
    (Materials_no, Max_iter,fobj,...
    dim,lb,ub,C3,C4);
13  figure,semilogy(Convergence_curve,'r')
14  xlim([0 1000]);
```

The function AOA has the following process: lines 27 to 35 update the density and volumes in the affected objects calculated to corresponding objects. Lines 19 to 25 aim to simulate the transference and density factor; in other words, this line transfers the condition from exploration to exploitation in time. Exploration phases represent the collision of the objects 29 to 35 at the same time as a factor to exploit the vector in line 33. When time passes, it is necessary to normalize the acceleration of objects introduced in the fluid; this is present in line 37. An essential phase is updating the body position in lines 39–59; on time, the fluid and body are stabilized, which means the best value has been found. Also is protected to not go out of boundaries in lines 61 to 78; it made an update the output data.

```
1   function [Xbest, Scorebest,Convergence_curve] =
    AOA(Materials_no,Max_iter,fobj, dim,lb,ub,C3,C4)
2   % Initialization
3   C1=2;C2=6;
4   u=.9;l=.1;    %paramters in Eq. (12)
5   X=lb+rand(Materials_no,dim)*(ub-lb);%initial
    positions Eq. (4)
6   den=rand(Materials_no,dim); % Eq. (5)
7   vol=rand(Materials_no,dim);
8   acc=lb+rand(Materials_no,dim)*(ub-lb);% Eq. (6)
9   for i=1:Materials_no
10      Y(i)=fobj(X(i,:));
11  end
12  [Scorebest, Score_index] = min(Y);
13  Xbest = X(Score_index,:);
14  den_best=den(Score_index,:);
15  vol_best=vol(Score_index,:);
16  acc_best=acc(Score_index,:);
17  acc_norm=acc;
18  for t = 1:Max_iter
19      TF=exp(((t-Max_iter)/(Max_iter)));    % Eq. (8)
20      if TF>1
21          TF=1;
22      end
23      d=exp((Max_iter-t)/Max_iter)-(t/Max_iter);    %
        Eq. (9)
```

```
24      acc=acc_norm;
25      r=rand();
26      for i=1:Materials_no
27         den(i,:)=den(i,:)+r*(den_best-den(i,:));
           % Eq. (7)
28        vol(i,:)=vol(i,:)+r*(vol_best-vol(i,:));
29        if TF<.45%collision
30              mr=randi(Materials_no);
31  acc_temp(i,:)=(den(mr,:)+(vol(mr,:)
    .*acc(mr,:)))./(rand*den(i,:).*vol(i,:));%Eq.(10)
32          else
33  acc_temp(i,:)=(den_best+(vol_best.*acc_best))./
    (rand*den(i,:).*vol(i,:));% Eq. (11)
34          end
35      end
36
37  acc_norm=((u*(acc_temp-min(acc_temp(:))))./
    (max(acc_temp(:))-min(acc_temp(:))))+l; % Eq. (12)
38
39      for i=1:Materials_no
40          if TF<.4
41              for j=1:size(X,2)
42              mrand=randi(Materials_no);
43             Xnew(i,j)=X(i,j)+C1*rand*acc_
               norm(i,j).*(X(mrand,j)-X(i,j))*d;%
               Eq. (13)
44              end
45          else
46              for j=1:size(X,2)
47                  p=2*rand-C4;  % Eq. (15)
48                  T=C3*TF;
49                  if T>1
50                      T=1;
51                  end
52                  if p<.5
53                    Xnew(i,j)=Xbest(j)+C2*rand*acc_
                      norm(i,j).*(T*Xbest(j)-
                      X(i,j))*d;  % Eq. (14)
54                  else
```

```
55                      Xnew(i,j)=Xbest(j)-C2*rand*acc_
                        norm(i,j).*(T*Xbest(j)-
                        X(i,j))*d;
56                  end
57              end
58          end
59      end
50
61      Xnew=fun_checkpositions(dim,Xnew,Materials_
        no,lb,ub);
62      for i=1:Materials_no
63          v=fobj( Xnew(i,:));
64          if v<Y(i)
65              X(i,:)=Xnew(i,:);
66              Y (i)=v;
67          end
68
69      end
70      [var_Ybest,var_index] = min(Y);
        Convergence_curve(t)=var_Ybest;
71      if var_Ybest<Scorebest
72          Scorebest=var_Ybest;
73          Score_index=var_index;
74          Xbest  = X(var_index,:);
75          den_best=den(Score_index,:);
76          vol_best=vol(Score_index,:);
77          acc_best=acc_norm(Score_index,:);
78      end
79
80  end  %end for
81
82  end   %end function
```

The fun_checkpositions function keeps the ranges within the allowed ranges in the problem set in the vectors.

```
1   function   vec_pos=fun_checkpositions(dim,vec_pos,
    var_no_group,lb,ub)
2   Lb=lb*ones(1,dim);
3   Ub=ub*ones(1,dim);
```

```
4   for i=1:var_no_group
5       isBelow1 = vec_pos(i,:) < Lb;
6       isAboveMax = (vec_pos(i,:) > Ub);
7       if isBelow1 == true
8           vec_pos(i,:) =Lb;
9       elseif find(isAboveMax== true)
10          vec_pos(i,:) = Ub;
11      end
12  end
13  end
```

Index